# Education, Communication and Decision Making on Renewable and Sustainable Energy

# Education, Communication and Decision Making on Renewable and Sustainable Energy

Editor

**Konstantinos Ioannou**

MDPI • Basel • Beijing • Wuhan • Barcelona • Belgrade • Manchester • Tokyo • Cluj • Tianjin

*Editor*
Konstantinos Ioannou
Hellenic Agricultural Organization DEMETER,
Forest Research Institute
Greece

*Editorial Office*
MDPI
St. Alban-Anlage 66
4052 Basel, Switzerland

This is a reprint of articles from the Special Issue published online in the open access journal *Sustainability* (ISSN 2071-1050) (available at: https://www.mdpi.com/journal/sustainability/special_issues/energy_education).

For citation purposes, cite each article independently as indicated on the article page online and as indicated below:

LastName, A.A.; LastName, B.B.; LastName, C.C. Article Title. *Journal Name* **Year**, *Article Number*, Page Range.

**ISBN 978-3-03936-589-0 (Hbk)**
**ISBN 978-3-03936-590-6 (PDF)**

© 2020 by the authors. Articles in this book are Open Access and distributed under the Creative Commons Attribution (CC BY) license, which allows users to download, copy and build upon published articles, as long as the author and publisher are properly credited, which ensures maximum dissemination and a wider impact of our publications.

The book as a whole is distributed by MDPI under the terms and conditions of the Creative Commons license CC BY-NC-ND.

# Contents

**About the Editor** .................................................................... vii

**Konstantinos Ioannou**
Education, Communication and Decision-Making on Renewable and Sustainable Energy
Reprinted from: *Sustainability* **2019**, *11*, 5262, doi:10.3390/su11195262 ................ 1

**Kerstin Tews**
The Crash of a Policy Pilot to Legally Define Community Energy. Evidence from the German Auction Scheme
Reprinted from: *Sustainability* **2018**, *10*, 3397, doi:10.3390/su10103397 ................ 5

**Daniel Efurosibina Attoye, Timothy O. Adekunle, Kheira Anissa Tabet Aoul, Ahmed Hassan and Samuel Osekafore Attoye**
A Conceptual Framework for a Building Integrated Photovoltaics (BIPV) Educative-Communication Approach
Reprinted from: *Sustainability* **2018**, *10*, 3781, doi:10.3390/su10103781 ................ 17

**Eleni Zafeiriou, Ioannis Mallidis, Konstantinos Galanopoulos and Garyfallos Arabatzis**
Greenhouse Gas Emissions and Economic Performance in EU Agriculture: An Empirical Study in a Non-Linear Framework
Reprinted from: *Sustainability* **2018**, *10*, 3837, doi:10.3390/su10113837 ................ 39

**Zacharoula Andreopoulou and Christiana Koliouska**
Benchmarking Internet Promotion of Renewable Energy Enterprises: Is Sustainability Present?
Reprinted from: *Sustainability* **2018**, *10*, 4187, doi:10.3390/su10114187 ................ 57

**Anna Mróz, Iwona Ocetkiewicz and Katarzyna Walotek-Ściańska**
Environmental Protection in School Curricula: Polish Context
Reprinted from: *Sustainability* **2018**, *10*, 4558, doi:10.3390/su10124558 ................ 69

**Dimitrios Drosos, Michalis Skordoulis, Garyfallos Arabatzis, Nikos Tsotsolas and Spyros Galatsidas**
Measuring Industrial Customer Satisfaction: The Case of the Natural Gas Market in Greece
Reprinted from: *Sustainability* **2019**, *11*, 1905, doi:10.3390/su11071905 ................ 91

**Sofia-Despoina Papadopoulou, Niki Kalaitzoglou, Maria Psarra, Sideri Lefkeli, Evangelia Karasmanaki and Georgios Tsantopoulos**
Addressing Energy Poverty through Transitioning to a Carbon-Free Environment
Reprinted from: *Sustainability* **2019**, *11*, 2634, doi:10.3390/su11092634 ................ 107

**A.H.T. Shyam Kularathna, Sayaka Suda, Ken Takagi and Shigeru Tabeta**
Evaluation of Co-Existence Options of Marine Renewable Energy Projects in Japan
Reprinted from: *Sustainability* **2019**, *11*, 2840, doi:10.3390/su11102840 ................ 125

**Stavros Tsiantikoudis, Eleni Zafeiriou, Grigorios Kyriakopoulos and Garyfallos Arabatzis**
Revising the Environmental Kuznets Curve for Deforestation: An Empirical Study for Bulgaria
Reprinted from: *Sustainability* **2019**, *11*, 4364, doi:10.3390/su11164364 ................ 151

**Evangelia Karasmanaki, Spyridon Galatsidas and Georgios Tsantopoulos**
An Investigation of Factors Affecting the Willingness to Invest in Renewables among Environmental Students: A Logistic Regression Approach
Reprinted from: *Sustainability* **2019**, *11*, 5012, doi:10.3390/su11185012 ................ 167

# About the Editor

**Konstantinos Ioannou** is a researcher of Forest Informatics at the Hellenic Agricultural Organization "DEMETER", Forest Research Institute. He is a graduate of the Department of Forestry and Natural Environment of the Aristotle University of Thessaloniki (A.U.Th). He holds a master's degree and a doctorate in Forest Informatics from A.U.Th and has received two scholarships for post-doctoral research. Additionally, he has received a research grant from the Stavros Niarchos Foundation for the implementation of the research project AgroComp. He is the author of more than 80 research papers in peer-reviewed journals and conferences and the holder of one national patent. He is also a reviewer for more than 20 Greek and international scientific journals, and he has participated in 15 research projects funded by the EU. His research interests focus on the broader field of artificial intelligence, developing decision-making systems, expert systems, artificial neural networks, and genetic algorithms, with the main goal of detecting future environmental problems and addressing them in a timely manner.

*Editorial*

# Education, Communication and Decision-Making on Renewable and Sustainable Energy

**Konstantinos Ioannou**

Hellenic Agricultural Organization DEMETER, Forest Research Institute, Vasilika, 57006 Thessaloniki, Greece; ioanko@fri.gr; Tel.: +30-2310461171-225

Received: 19 September 2019; Accepted: 22 September 2019; Published: 25 September 2019

**Abstract:** This editorial aims to introduce the themes and approaches covered in this special issue on education, communication, and decision-making on renewable and sustainable energy. At first, I discuss the themes and topics that have informed the creation of this special issue. Then, I provide an overview of the content of each paper that is included on the special issue. Additionally, this editorial provides a solid background on the relationships between the factors affecting decision-making on renewable energy sources as well as on the degree of influence education and communication takes part in the attitudes of the public towards renewable energy sources.

**Keywords:** decision-making; education; communication; investments; policy; RES

## 1. Introduction

During the last two decades, we have witnessed an evolution in the energy sector. Many countries throughout the world have been shifting their energy production methods from fossil fuel usage to more environmentally friendly methods. These methods are described under the term Renewable Energy Methods and propose the usage of Renewable Energy Sources (RES) based on wind, water, biomass, solar, and geothermal energy for the production of energy. This shift is mainly caused by the increase in public awareness on environmental problems and climate change, which are both related to the increase in Greenhouse Gas (GHG) emissions [1,2].

Alternate methodologies for reducing GHG emissions are also being applied. Energy saving is also an efficient way of confronting the problem. With the usage of the term "energy saving" we mean the reduction in the amount of energy consumed in a process or system, or by an organization or society, through economy, elimination of waste, and rational use. The application of initiatives regarding energy saving within school units can only bring benefits and lead towards reduction of energy cost [3]. Educational institutions are the most appropriate places in which students are taught energy conservation and involved in activities regarding rational energy management. Students are given opportunities to appreciate activities regarding energy saving and disseminate what they learnt in their wider social environment. The environmental education strategies applied constitute a significant educational process which strengthens student awareness of environmental issues [4].

The main goal of this special issue is the determination of methodologies which can be applied in education in order to raise the awareness of students as well as their families in issues related to renewable sources as well as in issued related to energy conservation [5,6].

Furthermore, an effort was made in order to determine the factors, parameters, and criteria affecting decision-making during the selection and investment in renewable energy sources [7].

Finally, an attempt was made in order to recognize methods for communicating the usage of RES and energy saving to the public. This is due to the fact that, in many communities, there are issues with the acceptance of RES installation as the public considers them as factors causing environmental degradation [8].

## 2. Overview of the Articles in the Special Issue

Zafeiriou et al. studied the relationship between agricultural carbon emissions equivalents and income per capita for the agricultural sector in different EU countries with the assistance of the nonlinear autoregressive distributed lag (NARDL) co-integration technique. Their findings validate the existence of a strong relationship between GHG emissions and agricultural income, since the co-integration among the two variables is established in all instances, while the asymmetric impact of agricultural income on carbon emissions may well provide policy makers with tools which, when implemented, may well promote the increase of agricultural income along with GHG effect mitigation in a successful way.

Attoye et al. aimed to develop a conceptual framework for an educative-communication approach for presenting BIPV proposals to encourage its adoption. The research paper focuses on developing a holistic research and market proposals which justify scholarly investigation and financial investment. By using a multiple case study investigation and Design Research Methodology (DRM) principles, the authors developed an approach which combines core communication requirements, the pillars of sustainability, and a hierarchical description of BIPV alongside its unique advantages. A two-step evaluation strategy involving an online pilot survey and a literature-based checklist was used to validate the effectiveness of the developed approach. The results show that understanding environmental and economic benefits are found to be significantly important to people who are likely adopters of BIPV ($p < 0.05$), making these benefits crucial drivers of adoption.

Kerstin Tews analyzed the effects of the privileges for "community energy actors" in the German auction scheme for on-shore wind energy. Those privileges aim to guarantee a level playing field for small actors and to enhance societal acceptance. The results of the first rounds of auctions did not merely reveal an acceptable level of losses due to recognized trade-offs between policy objectives. Instead, the results indicate a complete failure regarding all three objectives of the revised support scheme for renewables—controlled renewable energy expansion, actor plurality, and cost efficiency.

Andreopoulou and Koliouska evaluated the Renewable Energy Enterprises performance in the Internet in the Thessaloniki Prefecture regarding the characteristics of sustainability using a Multi-criteria Decision Analysis method called TOPSIS. The method was used to provide a ranking of the Renewable Energy Enterprises according to their sustainability. According to the results of the research, the Renewable Energy Enterprises achieve a good level of sustainability but not the optimum. However, it is suggested that the entrepreneurs should adopt modern environmental policy, sustainable marketing, green network framework, and a certified environmental management system in order to consider their enterprise sustainable.

Mroz et al. presented the results of a research on the integration of environmental protection issues into curricula by Polish teachers. In this research, it was assumed that the environmental protection issues included the challenges related to the sustainable management of natural resources. The sample consisted of 337 teachers of general subjects who were employed in schools in the Małopolska region (southern Poland) and working with students in lower-secondary (13–16 years old) and upper-secondary (16–20 years old) schools. The results of the research showed that many teachers know how to integrate environmental protection issues into their curricula.

Drosos et al. measured the industrial consumer satisfaction in the natural gas sector in Greece, by using the Multicriteria Satisfaction Analysis (MUSA) method. The researchers measured the industrial customer satisfaction based on criteria concerning the provided products and services, communication and collaboration with providers' staff, customer service, pricing policy, and website. The research results are based on the analysis of 95 questionnaires collected during the period between June 2017 and October 2017. The results show that the index of the global customer has a good performance as its value is about 74.99%.

Papadopoulou et al. investigated the views and attitudes of citizens of the Thessaloniki municipal area towards RES. For data collection, they used structured questionnaires which were filled out by performing personal interviews. Random sampling was performed to select the sample, and, in total, 420 citizens participated in the survey. The results showed that the respondents supported the replacement of lignite plants with renewable energy sources since they perceived that they constitute a necessary solution providing opportunities for economic growth and improvement to their quality of life. Finally, a vast majority of the responders expressed increased interest in future investment in photovoltaic systems, which, in their opinion, could contribute to improving air quality and increasing the energy independence, not only of Greece, but also of households.

Kularathna et al. evaluated the possible co-existence options available for Japan's MRE projects through data collected from interviews and questionnaire surveys in two development sites in Nagasaki and Kitakyushu in Southern Japan. The authors overcame the limitations of data unavailability and uncertainty by using the Dempster Shafer Analytic Hierarchy Process (DS-AHP) for evaluating the best co-existence strategy out of five potential options. The results indicate that local fisheries prefer the oceanographic information sharing option. whereas most of the other stakeholders prefer using local resources to construct and operate the power plant, creating business involvement opportunities for the local community.

Tsiantikoudis et al. studied the economic growth—environmental degradation relationship—namely, the environmental Kuznets curve (EKC) hypothesis—in alignment with the autoregressive distributed lag (ARDL) approach. The novelty of the study is attributed to the usage of the carbon emissions equivalent deriving by deforestation as an index for environmental degradation. In addition, the researchers used the gross domestic product (GDP) per capita as a proxy for income, being determined as an independent variable. The entire research was performed for Bulgaria, a country which recently joined the European Union. Research findings cannot validate the inverted U-shape of the EKC hypothesis; instead, an inverted N pattern was confirmed.

Karasmanaki et al. tried to identify the most important factors that affect environmental students' willingness to invest in renewable energy by developing a logistic regression model. According to their analysis, the results showed that the majority of the participants expressed their willingness to invest in RES. The most important factors determining this willingness were the environmental values, the low risk and profitability of renewable investments as well as the preference for certain energy types. However, willingness to invest was irrespective of the current taxation and subsidies, suggesting that significant improvements are required in these areas.

**Funding:** This research received no external funding.

**Acknowledgments:** I would like to acknowledge the support of the authors and reviewers who have contributed to this special issue, to whom we express our sincere thanks.

**Conflicts of Interest:** The author declares no conflict of interest.

## References

1. Konstantinos, I.; Georgios, T.; Garyfallos, A.; Zacharoula, A.; Eleni, Z. A spatial decision support system framework for the evaluation of biomass energy production locations: Case study in the regional unit of drama, Greece. *Sustainability* **2018**, *10*, 531.
2. Konstantinos, I.; Georgios, T.; Garyfalos, A. A Decision Support System methodology for selecting wind farm installation locations using AHP and TOPSIS: Case study in Eastern Macedonia and Thrace region, Greece. *Energy Policy* **2019**, *132*, 232–246. [CrossRef]
3. Castleberry, B.; Gliedt, T.; Greene, J.S. Assessing drivers and barriers of energy-saving measures in Oklahoma's public schools. *Energy Policy* **2016**, *88*, 216–228. [CrossRef]
4. Simsekli, Y. An Implementation to Raise Environmental Awareness of Elementary Education Students. *Procedia Soc. Behav. Sci.* **2015**, *191*, 222–226. [CrossRef]
5. Petkou, D.; Tsantopoulos, G.; Tampakis, S.; Panagiotou, N. Typology of teachers based on their attitudes and behaviours as shaped by the influence of mass media on environmental issues. *J. Environ. Prot. Ecol.* **2018**, *19*, 1352–1361.
6. Lefkeli, S.; Manolas, E.; Ioannou, K.; Tsantopoulos, G. Socio-cultural impact of energy saving: Studying the behaviour of elementary school students in Greece. *Sustainability* **2018**, *10*, 737. [CrossRef]

7. Ioannou, K.; Lefakis, P.; Arabatzis, G. Development of a decision support system for the study of an area after the occurrence of forest fire. *Int. J. Sustain. Soc.* **2011**, *3*, 5–32. [CrossRef]
8. Tampakis, S.; Arabatzis, G.; Tsantopoulos, G.; Rerras, I. Citizens' views on electricity use, savings and production from renewable energy sources: A case study from a Greek island. *Renew. Sustain. Energy Rev.* **2017**, *79*, 39–49. [CrossRef]

© 2019 by the author. Licensee MDPI, Basel, Switzerland. This article is an open access article distributed under the terms and conditions of the Creative Commons Attribution (CC BY) license (http://creativecommons.org/licenses/by/4.0/).

Article

# The Crash of a Policy Pilot to Legally Define Community Energy. Evidence from the German Auction Scheme

Kerstin Tews [1,2]

[1] Environmental Policy Research Centre, Freie Universität Berlin, 14195 Berlin, Germany; kerstin.tews@fu-berlin.de; Tel.: +49-30-838-55098
[2] Bavarian School of Public Policy, Technical University Munich, 80333 Munich, Germany; kerstin.tews@hfp.tum.de

Received: 29 August 2018; Accepted: 21 September 2018; Published: 24 September 2018

**Abstract:** "Community energy" is a highly contested issue not only in the German energy transition governance but also in the recent legislative procedure to recast energy market legislation within the EU's "Winter Package". This paper analyses the effects of the privileges for "community energy actors" in the German auction scheme for on-shore wind energy. Those privileges aim to guarantee a level playing field for small actors and to enhance societal acceptance. The results of the first rounds of auctions did not merely reveal an acceptable level of losses due to recognized trade-offs between policy objectives. Instead, the results indicate a complete failure regarding all three objectives of the revised support scheme for renewables—controlled renewable energy expansion, actor plurality and cost efficiency. The paper discusses whether the policy motivations translate appropriately into legislation. It suggests differentiating clearly the economic risks for small actors from the added value that is attributed to community energy actors. De-coupling these—often mixed—motives behind the demands for actor plurality unveils policy approaches that more adequately fit with these distinct motives. The paper finally proposes rather to integrate the politically desired values into the bid evaluation criteria instead of granting privileges to specific actors that are assumed to provide these values.

**Keywords:** renewable energy; governance; community energy; citizens' energy companies; actor plurality; acceptance; energy transition; auction scheme; Germany

---

## 1. Introduction

"Community energy" is a highly contested issue not only in the German energy transition governance but also in the recent legislative procedure to recast energy market legislation within the EU's "Winter Package" [1].

In Germany, "community energy actors" significantly pushed the deployment of renewable energies (RE) in the past. In the course of the shift of the support scheme for renewable energies from a price-based feed-in tariff to a volume-based competitive auction, the German Government for the first time legally defined the term "citizens' energy companies". This term was officially established in order to select those energy actors who were privileged by special auction rules. These special rules were justified by the policy objectives to enhance actor plurality and societal acceptance of the energy transition.

The paper analyses effects of the German provisions to privilege "community energy actors" against the background of the objectives of the revised German renewable energy policy. It finds that the results of these provisions for community energy actors in the German auction scheme do not just represent an acceptable level of losses due to recognized trade-offs between the three main objectives—controlled RE expansion, actor plurality and cost efficiency—but instead a complete failure with regard to all three of these objectives.

The paper takes a closer look at the motivations that officially underpinned these special provisions for citizens' energy actors in order to draw lessons from that policy pilot. These lessons might also help other European countries implementing the envisaged EU energy market legislation in the future. The EU-Parliament's version of the recast renewable energy directive suggests that member states have "[ ... ] to put in place an enabling framework to promote and facilitate participation by renewable energy communities in the generation, consumption, storage and sale of renewable energy" [2] (article 22(2a)).

The paper suggests to clearly distinguish between the risks for small actors to take part in auctions from the added societal value, attributed to the energy related activities of these actors. This approach will help to define the necessary policy design elements that are suitable to enable small and community energy actors to participate in energy market activities, to provide the politically desired benefits, and to prevent disastrous policy failures similar to those in the German case.

## 2. "Community Energy" in the German Energy Policy Context

*2.1. Occasion: Instrumental Shift in the Support Scheme for Renewables*

The term "community energy" or "citizens' energy" entered the political agenda in the course of the debate on the shift of the national support scheme for renewable energy. In 2014, the German government decided to switch from a price-based to a volume-based support scheme. This fundamental instrumental shift was "forced" by external pressure (EU-Commission's State Aid Guidelines [3]) on the one side, but also by domestic debates about the affordability of the energy transition and the increasing costs of RE deployment on the other side (for more details on drivers and implications of this instrumental shift, see [4]).

The price-based support scheme for RE—the so-called feed-in-tariff—functioned as a shelter, allowing small-scale renewable electricity producers to develop in a niche. These new actors have challenged established patterns of domestic energy policy interaction through experimentation and innovation at a decentralized level [5]. According to a survey carried out by trend:research GmbH and the Leuphana Universität Lüneburg [6] nearly half (46.6 percent) of the total RE capacity installed in Germany was owned by citizens and collective citizens' energy initiatives before the introduction of the auction scheme. It has been argued by many scientists and proponents of this "bottom-up" energy transition that these new energy actors are not purely driven by maximum profit-seeking motives (e.g., [7]). Instead, they combine their engagement in energy business with a common good orientation in terms of local development, inclusive democracy, citizens' engagement and social innovation.

*2.2. Risk of Auction Schemes and Measures of the German Government to Counter These Risks*

Many stakeholders perceived the instrumental shift as a serious threat for a further engagement of these new energy actors who have driven the transition in Germany thus far. Various empirical studies pointed to the risks of auction schemes (i.e., [8,9]). Auctions would disadvantage local small-scale investors, as they are less able to diversify risks related to the uncertainty of successful bids and to cover higher transaction costs associated with the participation in auctions. In addition to the loss of actor plurality, these studies have pointed to various other risks. They comprise the threat of spatial concentration of generation facilities (hotspots) and the respective high burdens on the grid infra-structure, innovation barriers due to the exclusion of less mature RE technologies as well as the risks of low actual project implementation rates, as demonstrated by the experiences of auction schemes in other countries. Low implementation rates would in fact threaten the main purpose of a volume-based and competitive coordination mode for RE development: the cost-effective achievement of the politically determined annual RE expansion targets.

These other risks were discussed during consultations and the German government introduced various legal provisions into the Renewable Energy Act (EEG) to counter these risks:

(a) The "reference yield model", for example, was adapted in order to prevent hotspots in particularly wind-intensive geographical areas. It aims to enable wind energy plants to operate throughout the federal territory by remunerating at different rates, depending on location. However, the actual bid in a given tender has to refer to an administratively defined reference site in order to make the calculated prices of the competing bids comparable.

(b) The German government additionally opted for technology specific tenders to counter the risk of blocking further development of (less mature) RE technology and to safeguard diversification of the RE sources. The recent results of the first pilot auction round combining wind and PV verified the relevance of this approach. All awards in this combined pilot tender went to PV-based bids [10].

(c) The core approach to handle the risk of low implementation rates and, thus, to miss the expansion targets was the concept of "late" auctions. "Late" refers to the point in time of the development process of a given wind energy project. As a rule, the construction permit (grant of approval pursuant to the Federal Immission Control Act, BImSchG) is a central pre-qualification to participate in a tender. Granting this construction permit is a complex process, as all potential environmental impacts of the specific planned wind turbine/park have to be considered. This process requires time, knowledge and capacities, i.e., it induces a high amount of transaction costs. Having passed this process successfully can be assessed as an indicator for a high propensity of actual implementation of the awarded bids. The short implementation deadlines of maximum 2.5 years aim additionally at achieving expansion targets without delay.

Particularly, the knowledge about the political intention of conceptualizing "late" auctions, i.e., to guarantee high and timely implementation rates, also helps to understand the implicit assumptions underpinning the special regulation for citizens' energy companies. It will also help us to evaluate the impact of that special regulation according to the empirical results of the auction schemes in Germany as will be presented in the following sections.

### 2.3. Legal Definition and Special Provisions for Citizens' Energy Companies

During the debate on the revision of the EEG, the German government repeatedly declared that it will pursue actor plurality in its future energy transition efforts and will not threaten regional and local efforts towards a low-carbon energy transition. The results of the previous pilot bidding rounds on ground mounted photovoltaic systems (2014–2016) verified the concerns regarding a loss of actor plurality and the exclusion of small players. The majority of the capacity awarded went to bidders with more than one bid and bidders who feature intercompany ties with other successful bidders (see [5]). Confronted with these results the government introduced special regulations in order to create a level playing field for local citizen-based energy companies related to wind specific tenders.

#### 2.3.1. Definition of "Citizens' Energy"

The revised EEG—adopted in 2016 and entered into force in January 2017—for the first time legally defined the term "citizens' energy".

§ 3 EEG 2017 defines a citizens' energy company as an entity,

- "which consists of at least ten natural persons with voting right
- in which at least 51 percent of the voting rights are held by natural persons which live in the urban or rural district in which the onshore wind energy installation is to be erected,
- in which no member or shareholder of the undertaking holds more than 10 percent of the voting rights of the undertaking" [11].

§ 36g EEG 2017 furthermore defines, that

- members or shareholders are not allowed to have concluded contracts to transfer shares or voting rights in order to circumvent provision of § 3;
- in the case of a successful bid, a 10 percent financial stake has to be offered to the municipality, where the installation is erected;
- The scale of the bid is restricted: up to six wind turbines or a maximum capacity of 18 MW (ibid.).

However, community energy or citizens' energy is not a uniform phenomenon. Literature offers a variety of terms, characteristics, narratives which represent different theoretical approaches and ideational concepts (e.g., [12] (p. 897)). The German Government's approach to legally define those actors that are eligible to benefit from the special legal provisions tried to operationalize particularly the following characteristics:

(a) small scale nature and limited number of projects;
(b) citizens' control and
(c) locality of investments and returns.

The motivations underlying this set of criteria were to counteract small players' risk of having to bear up-front costs without guarantees to win auctions, and their limited opportunity to di-versify this risk through multiple projects. An additional and related motivation was to buttress acceptance of the local population via the local embeddedness of investors. Thus, the target of the special provisions was to create a level playing field for these actors in an imperfect market mainly for competition concerns.

Furthermore, successful bids by citizens' energy companies are awarded according to the uniform pricing principle in contrast to pay-as-bid principle for "ordinary" bidders. The uniform pricing principle means that the value of the award for bids of citizens energy companies shall be the value of the award of the highest bid awarded in the respective auction round.

2.3.2. Special Auctioning Rules for Citizens' Energy Companies

§ 36g of the EEG 2017 defines those special rules, which solely apply to citizens' energy companies (CEC). They are intended to provide the necessary level playing field for these actors to participate in auctions. The most relevant provisions comprise

(a) the allowance to submit a bid before the granting of the construction permit (pursuant to the Federal Immission Control Act (§ 36g(1)), and
(b) the allowance of longer realisation times for project implementation of up to 4.5 years compared to 2.5 years for "ordinary" bidders (§ 36g(2)).

These exceptional provisions can be perceived as *a fundamental derivation* from the core principle of the German auction scheme—the principle of "late" tenders. Having in mind the underpinning motivation of "late" tenders—safeguarding target achievement—it becomes clear that the legislator implicitly assumed citizens' energy as a rather small segment among the potential bidders.

## 3. Impact of Special Regulation for Citizens' Energy Companies

*3.1. Success of Citizens' Energy?*

In contrast to any expectation and to many concerns of citizens' energy advocates about potential shortcomings of the EEG's provisions for citizens' energy, the results of the first auction in May 2017 surprised with an overwhelming success of bidders that made use of the special rules for citizens' energy companies. The privileges for CEC, conceptualized as exceptions, became the rule in the market "game".

Shocked by this unintended result which seriously threatened the achievement of the planned RE expansion and climate targets, the German Bundestag immediately announced a moratorium regarding § 36g(1) and stopped the possibility for CEC to offer a bid before the construction permit has been granted. However, this moratorium took force only for the fourth and fifth auction rounds in 2018, before a final adaptation of the EEG will be implemented. Consequently, also in the second and third auction round in 2017, citizens' energy companies won almost all of the awards (see Figure 1).

How to interpret these results? Do they represent the legislator's failure to assess ex-ante the strength and volume of the citizens' energy segment? Or, did the legal definition not suffice to address the intended actors—citizens which initiate on their own wind energy projects in their neighborhood? Or, is the privilege offered to citizens' energy companies so economically attractive that even professional energy actors altered their projects into projects which fit with the legal definition of a CEC-project? Or is there something inconsistent within the whole policy design?

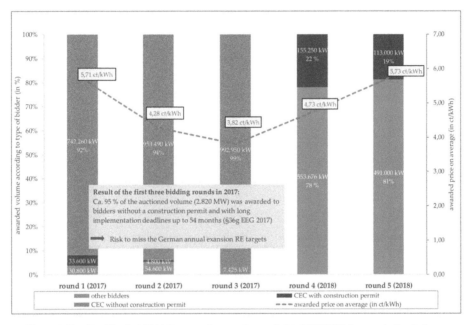

**Figure 1.** Results of the first 5 bidding rounds on onshore wind in 2017/2018; Source: Author's figure based on data from Bundesnetzagentur [13].

*3.2. CEC as Business Model for Rational Economic Actors*

The press release of the Federal Network Agency (Bundesnetzagentur) regarding the results of the first auction round in May 2017 [14] rather neutrally reported the success of citizens' energy. The press release following the second bidding round already revealed the Agency's awareness about the character of the successful CEC bidders: "The majority of the awards for bids of citizens' energy companies go to companies whose bids make it clear that they are at least organisationally assigned to a single project developer" (author's translation of the press release, [15]).

The single project developer mentioned in the press release—the Saxonian UKA-Group—is ac-cording to data from 2016 the second biggest developer of on-shore wind energy project in Germany. In an interview for the magazine "Erneuerbare Energien" in November 2017 the company's managing partner, Gernot Gauglitz, publicly explained the company's strategy as the most adaptive business model to comply with the induced competitive pressure in the German auction scheme [16] According to him, UKA was forced to emulate the behaviour of other project developers that won in the first bidding round, where UKA failed to win any of its "ordinary" bids. The adaptation of its strategy was so successful that UKA-connected CEC dominated the second bidding round with a 68 percent share of the total awarded volume (see Table 1).

Table 1. Results of the second bidding round in 2017: A single player's dominance.

| Total Number of Awards | Awards for CEC | Awards for UKA-Connected CEC-Bidders | Share UKA of Total Number of Awards |
|---|---|---|---|
| 67 | 60 | 37 | 55% |
| Total awarded volume 1013 MW | Awarded volume for CEC 958 MW | Awarded volume for UKA-connected CEC-bidders 690 MW | Share UKA of total awarded volume 68% |

Source: Author's compilation based on [13,15,16].

In sum, the successful bidders privileged as CEC were set up by a very small number of professional project developers, who do not act as shareholders or members of the CEC—according to the legal definition—but as general contractors or service providers. A brief analysis of data of the Bundesnetzagentur [13] and the online-trade register [17] about the winners of the second round revealed that a lot of the successful CEC were formally founded just a few days before the auction deadline. They also have similar names, for example "Umweltgerechte Bürgerenergie", and—although the planned erection site is located in different municipalities—the registered office of those formally distinct CEC is the same and situated in the Saxonian town Meißen, identical with the postal address of UKA.

In fact, those project developers, who transformed their projects into CEC-projects made use of the privileges for CEC:

- to take part in the bidding procedure without having a construction permit, and
- the longer time span between awards and realisation deadline (up to 4.5 years).

These provisions offered opportunities for underbidding other "ordinary" bidders, a well–known phenomenon of competitive auction schemes. The exceptions—intended to support a small segment of rather unprofessional citizen-based projects—were used by typical business actors in an economically rational way. They calculated their expected returns (bids) with more effective wind turbines which according to their market analyses will only be developed in the near future. Thus, these special provisions for CEC have been used by professional actors to an extent that devaluated all existing construction permits for wind energy projects, as these permissions always have to relate to a specific type of an available wind turbine technology [18] (p. 4).

In sum, it can be stated that professional project developers, like UKA, have carefully analysed the market as well as the political framework conditions, and have rationally adapted their business model. Although diverse media—such as the German newspaper "Die Welt"—criticized "the dirty tricks with citizens' energy" [19] (author's translation), the CEC connected with these project developers meet all legal requirements of a CEC. Thus, the expressed complaint, it would be "fake" CEC that won the awards does suggest a knowledge, what "real" CEC is. Unfortunately, "real" CEC is not clearly defined—neither in scientific literature nor in the definition of citizens' energy in the EEG.

*3.3. Evaluation: Total Failure Regarding the Intended Triangle of Objectives*

As almost 95% of the auctioned volume in 2017 was awarded to bidders without a construction permit, it is doubtful how many of these projects will actually be realised. This results in uncertainty whether one core element of the instrumental shift—the volume-based approach to steer RE expansion—can be actually achieved. Additionally, the longer implementation deadlines of up to 54 months for these projects indicate that most of them—if ever—will not be realised before the year 2020. According to the Green Party in the German parliament, this would cause a sharp decline of the annual expansion of wind energy in Germany to 1500 MW or less in 2019 and 2020 compared with 5000 MW per year in 2016 and 2017. Moreover, this would imply missing the German and the mandatory European renewable energy targets [20].

The wind energy industry is complaining tremendously about these results as they represent a serious threat to their business model: uncertainties regarding the actual demand of wind turbines and/or the expected delay of the installation of wind turbines that is assumed to create a serious investment gap for the wind energy industry [18] (p. 4).

As explained above, the auctions were not able to foster actor plurality, the second core objective of the specific provisions, either. Instead, a small number of professional project developers dominated.

What about cost efficiency, the third objective of the German volume-based auction scheme and the core argument of proponents of competitive approaches to determine the price for RE?

At first glance, Figure 1 obviously illustrates a considerable decrease in the average price per kWh of wind energy in the first three bidding rounds, from 5.71 Ct/kWh to 3.82 Ct/kWh. The remuneration that would have been paid according to the previous administratively fixed feed-in-tariff (EEG 2014) would be—nominally averaged—6.16 Ct/kWh for plants erected at the beginning of 2018 [21]. Thus, it was claimed by the government that this core objective was reached by the German auction scheme.

However, the price results presented in Figure 1 are determined by the calculated prices of the bids. They do not show the actual remuneration according to the wind situation at the erection site as defined in the above-mentioned reference yield model. The awarded projects of the auctions in 2017 are planned to be erected at locations of a middle wind quality of 90% (first bidding round) or 85% (second and third bidding round) of the wind quality at the standard reference side [21]. That means that actual remuneration payments will be higher than prices calculated for the bids.

Researchers of the German Fraunhofer Institute IEE and IZES have questioned the alleged cost-cutting effect [21]. They simulated a comparison of remunerations based on the previous price-based support scheme and the results of the current volume-based support scheme by considering the reference yield model—i.e., the wind-quality at the erection site of the awarded CEC projects in the first three bidding rounds in 2017. They discovered up to one-fourth higher costs of the awards for CEC projects in the first bidding round compared to the previous remuneration based on the feed-in tariff scheme. Only the results of the third bidding round have shown a 10 percent reduction of the level of remuneration compared to the previous system [21]. Keeping in mind that these prices were a result of the very specific underbidding approach of successful bidders incentivized by the special provisions of § 36g, the magnitude of this decrease can be seriously questioned. The subsequent increase in the average prices per kWh after the moratorium took force in 2018 confirms this assessment (see Figure 1). Energy experts had expected this price increase [21]. All bidders with construction permits, including the unsuccessful bidders of the first three rounds, had to calculate their bids based on the available wind technology and no longer needed to adjust their bids to the expected underbidding strategy of privileged bidders without permits.

## 4. Discussion

What kind of lessons can we draw from this severe policy failure to meet the intended objectives?

The German legislator has integrated secondary policy objectives—a geographically more balanced distribution of wind energy plants and actor plurality—into an economic instrument, which mainly addresses the economic efficiency of the price building process. A trade-off with the economic efficiency objective of the auction scheme was—according to the legislator's justification of the German auction design [22] (p. 147)—politically accepted. Such trade-offs are quite common outcomes of political processes, whose protagonists have to find compromises between interdependent policy objectives.

*4.1. Inadequate Integration of Secondary Objectives*

However, the results of the German auction scheme do not represent a more or less acceptable level of assumed losses due to recognized trade-offs between the three main objectives of controlled

RE expansion, actor plurality and cost efficiency. Instead, we see a complete failure with regard to all of these three objectives (see above).

There might be economists that feel confirmed in their assessment that any intervention into a market would be a market distortion and should therefore be avoided. However, the renewable energy market would not exist without state intervention and a piecemeal criticism of (presumed) market distortion would be the wrong lesson from the German "disastrous" policy pilot to create a level playing field for new citizens-based energy actors.

A closer look at the policy motivation underlying the definition of citizens' energy companies and the relation between these motivations and their operationalisation might give us more insights to draw adequate lessons.

Integrating the objective of "actor plurality" into the policy design of the German auctions scheme was motivated by, first, purely competition concerns to prevent market dominance of a few big energy actors [22]. This challenge was met by general provisions that may counteract the risks of small actors—e.g., a generalized de-minimis threshold of 750 kW. However, the German legislator additionally defined an enabling policy framework for certain small actors to participate in wind energy auctions. The sole argument to explain that specific focus on citizen-based energy companies can be derived from the official justification of § 36g EEG 2017: "In particular, locally anchored citizen energy companies have made a significant contribution to the necessary acceptance of new on-shore wind energy projects. Without this acceptance, the expansion of wind energy cannot be achieved in the planned amount" (author's translation, [22] (p. 217)).

However, if citizens' energy is perceived by the legislator as the provider of the necessary acceptance of wind energy expansion, then it seems to be inconsequential, that the citizens' energy segment was implicitly conceptualized as just a small segment among the potential bidders. To reiterate, all the special rules for CEC are a fundamental derivation from the core principle of the German "late" auction design (see above).

*4.2. The Need to Decouple Multiple Motives*

A possible approach to learn from these failures is to distinguish clearly between the two motivations underlying the objective "actor plurality". The first one addresses the risks of small actors, and the second one addresses the added value provided by citizen energy actors. De-coupling these two motivations might be helpful to find an adequate policy design.

4.2.1. Addressing the Risks of Small Actors

If design elements are supposed to help to safeguard market access for small actors for com-petition concerns, then the very specific risks of small actors must be addressed. This could be realized either with support measures outside of the auction, e.g., by providing counselling during preparation of a bid, or by exempting small actors from auctions altogether. Currently, the German generalized de-minimis threshold for exemptions from competitive tendering is at 750 kW. However, this threshold has in fact no relevance for wind energy projects, as the market standard of available wind turbines is above this threshold at 2.5 to 3 MW on average. The German Government rejected to use the full room for maneuver to exempt smaller projects from tendering which has been given by the European Commission state aid guidelines [3]. According to the EU Competition Commissioner's clarification from January 2016 about the misleading formulation of the de-minimis rule for wind power in the state aid guidelines, there is a possibility of "exemption from the competitive tendering requirement […] for wind projects with a maximum of 18 MW installed power" [23] (author's translation of the answer of Competition Commissioner Margrethe Vestager to the German Federal Wind Association).

### 4.2.2. Addressing the Added Value Attributed to Community Energy Actors

However, if acceptance is the determining criterion, then we need to look in more detail at what in fact contributes to societal acceptance, as the legislator did not specify sufficiently, why citizen energy actors have significantly contributed to this objective. The items "majority of natural persons" of the members or shareholders of the CEC and their "place of residence" in the district of the planned wind energy plant have been used to operationalize characteristics of citizen-based energy actors and local embeddedness. Both characteristics are often linked in the literature with acceptance issues (e.g., [24]).

Yet, it is not too difficult to imagine that ten individual citizens who invest in a joint local economic undertaking can also act in a purely egoistic manner, i.e., without taking into account economic welfare or social acceptance in their neighborhood in a more caring manner than other economic actors. The acceptance issue, thus, was not operationalized, but only attributed a priori to community energy actors.

The acceptance issue is also a core objective of the EU's so-called Winter Package "Clean Energy for All Europeans" [1]. Currently the Recast Renewable Energy Directives and the Recast Electricity Market Directive are subject of the Trilogue between Commission, Council and Parliament. Both recast directives include definitions and special provisions for local (renewable) energy communities. Whereas the Commission and the Parliament strongly favor an enabling framework for community energy actors in their positions concerning theses directives, the Council's amendments to the Commission's proposals [25] indeed rather pronounce the costs induced by community energy actors and their financial responsibility " ... for the imbalances they cause in the system" [26] (p. 62).

The European Commission (EC), assisted by the European Parliament, has "discovered" the added value that local community energy can provide. In its proposal for the Recast Electricity Directive the EC defines local energy communities (LEC) in the following way (article2(6)): "Local energy community means: an association, a cooperative, a partnership, a nonprofit organisation or other legal entity which is *effectively controlled* by local shareholders or members, *generally value rather than profit-driven*, involved in distributed generation and in performing activities of a distribution system operator, supplier or aggregator at local level, including across borders." [25] (p. 52, author's emphasis).

Thus, the Commission includes in its definition not only the various energy business activities a LEC can perform, but also declares "local control" and "value-driven" as key features making community energy actors distinct from other energy business actors. The added value for the European energy transformation process the EC attributes to LEC in its recitals for the recast electricity directive can be summarized as follows: LEC provide common goods: "Community energy [ ... ] help[s] fight energy poverty [ ... ] enables certain groups of household consumers to participate in the energy market [ ... ] Where they have been successfully operated such initiatives have delivered economic, social and environmental value to the community that goes beyond the mere benefits derived from the provision of energy services." [25] (recital 30).

## 5. Conclusions

Citizen energy actors combine their energy business activities with a broader common good orientation in terms of local development, inclusive democracy, citizens' engagement and social innovation. All of these contributions are increasingly needed to counteract the observable loss in acceptance and the growing strength of populism not only with regard to energy issues.

However, attributing these contributions in legislation to a rather narrowly defined group of actors such as in the German definition of citizens' energy companies does not seem to be adequate. It excludes a couple of well-known other decentralized initiatives as the municipality-based 100% Energy Regions or certain municipal utilities and other actors which include common good provisions into their energy business activities (see [5]). Moreover, it seems questionable whether the provision of those common goods that help raise acceptance can be realized via a definition of privileges for a defined category of actors. It will always be difficult—as evidenced by the German auctions—to find selective criteria to define the eligible actors that are not vulnerable for "abuse".

If acceptance is perceived as a result of the provision of those common goods as regional development, participation, the experience of self-efficacy as well as social inclusion of vulnerable people, then it seems to be more appropriate to change the bid evaluation criteria, i.e., the criteria defining how bids are awarded.

Currently, the bid selection in the German scheme works as a price-only selection process, comparable to the majority of auction schemes in the EU.

However, there are examples of multi-criteria approaches to select the winning bids in order to pursue multiple policy goals (e.g., [27]). In France for example, the bid evaluation in the PV auctions is based on the offered price (two thirds) and the environmental impact in the form of the panels' carbon footprint (one third). The German solar branch has repeatedly referred to the French approach as more suitable than the German price-only approach, because it better protects domestic producers of solar panels as they perform better with the environmental criteria then their competitors for non-European countries.

Another example is the South-African "Renewable Energy Independent Power Project Procurement Program" that strongly relies on non-price factors in bid evaluation. The bid price counts for 70 percent, whereas the remaining 30 percent are given to a "composite score covering job creation, local content, ownership, management control, preferential procurement, enterprise development and socioeconomic development" [28].

Including these secondary objectives into the bid evaluation criteria means—of course—com-promising on the economic efficiency of auction schemes. It clearly increases costs for bidders and the regulator in evaluating these bids. However, in most of the countries where these secondary objectives have been applied in bid evaluation, they helped to promote social acceptability of RE expansion policy and local economic development [27] (p. 39).

Whether to exempt those projects which are assumed to provide such added value from competitive auctions altogether or whether to include these objectives into bid evaluation criteria, depends on a sophisticated cost-benefit analysis, which goes beyond the mere consideration of static cost efficiency. If the proclaimed "secondary" objectives of the German auction scheme are more than just lip service, then policy makers have to devote more attention to them.

Having in mind the increasing share of citizens who feel disconnected to their political elites in Germany as well as in other countries of the European Union—and energy transition mirrors that development—it becomes clear that addressing public welfare and social inclusion have to be key objectives of further energy transition effort.

The added value provided so far by community energy actors—including municipalities, cooperatives and other decentralized community-based actors—in the form of democratization of energy business, opening opportunities to experience social self-efficacy, local development and social inclusion should be honored by democratic policy makers facing growing voter's mistrust and populist political competitors.

It will not suffice to think only in terms of providing a level playing field in energy related economic activities in an imperfect market for purely competition concerns. Instead, there is a need to create a level playing field for common-good-oriented visions and paradigms on societal and energy transition trajectories.

**Funding:** This article is largely based on research conducted as part of the project "Impact of political framework conditions on citizens' engagement" at the Technical University Munich (TUM). The TUM-project is part of the working package "Change in Multi-Level Governance-Systems" of the KOPERNIKUS-Project "ENavi" funded by the German Ministry for Education and Research (BMBF). Furthermore, the author acknowledges support by the German Research Foundation and the Open Access Publication Fund of the Freie Universität Berlin.

**Acknowledgments:** The author would like to thank the three anonymous reviewers for their comments.

**Conflicts of Interest:** The author declares no conflict of interest.

## References

1. European Commission. Clean Energy for All Europeans. Available online: https://ec.europa.eu/energy/en/topics/energy-strategy-and-energy-union/clean-energy-all-europeans (accessed on 29 August 2018).
2. European Parliament. *Promotion of the Use of Energy from Renewable Sources*; Amendments Adopted by the European Parliament on 17 January 2018 on the Proposal for a Directive of the European Parliament and of the Council on the Promotion of the Use of Energy from Renewable Sources (Recast) (COM(2016)0767-C8-0500/2016—2016/0382(COD); P8_TA-PROV(2018)0009; European Parliament: Brussels, Belgium, 2018.
3. European Commission. *Communication from the Commission: Guidelines on State Aid for Environmental Protection and Energy 2014–2020 (2014/C 200/01)*; European Commission: Brussels, Belgium, 2018.
4. Tews, K. Europeanization of Energy and Climate Policy: The Struggle Between Competing Ideas of Coordinating Energy Transitions. *J. Environ. Dev.* **2015**, *24*, 267–291. [CrossRef]
5. Beermann, J.; Tews, K. Decentralised laboratories in the German energy transition. Why local renewable energy initiatives must reinvent themselves. *J. Clean. Prod.* **2017**, *169*, 125–134. [CrossRef]
6. Holstenkamp, L. *Definition und Marktanalyse von Bürgerenergie in Deutschland*; Research and Leuphana Universität Lüneburg: Lüneburg, Germany, 2013.
7. Holstenkamp, L.; Kahla, F.; Degenhart, H. Finanzwirtschaftliche Annäherungen an das Phänomen Bürgerbeteiligung. In *Handbuch Energiewende und Partizipation*; Holstenkamp, L., Radtke, J., Eds.; Springer: Wiesbaden, Germany, 2018; pp. 281–302.
8. Ecofys. *Design Features of Support Schemes for Renewable Electricity*; A Report Compiled within the European Project "Cooperation between EU MS under the Renewable Energy Directive and Interaction with Support Schemes"; European Commission, DG ENER: Utrecht, The Netherlands, 2014.
9. Grashof, K.; Kochems, J.; Klann, U. *Charakterisierung und Chancen kleiner Akteure bei der Ausschreibung für Windenergie an Land*; Fachagentur Wind an Land e.V.: Berlin, Germany, 2015.
10. Bundesnetzagentur. Pressemitteilung. *Ergebnisse der Gemeinsamen Ausschreibung von Wind- und Solaranlagen*; Bundesnetzagentur für Elektrizität, Gas, Telekommunikation, Post und Eisenbahnen: Bonn, Germany, 2018.
11. BMWI. Informal English Version of the EEG 2017. Available online: https://www.bmwi.de/Redaktion/DE/Downloads/E/eeg-2017-gesetz-en.pdf?__blob=publicationFile&v=8 (accessed on 20 April 2018).
12. Holstenkamp, L. Einleitende Anmerkungen zum Ländervergleich: Definition von Bürgerenergie, Länderauswahl und Überblick über Fördermechanismen. In *Handbuch Energiewende und Partizipation*; Holstenkamp, L., Radtke, J., Eds.; Springer: Wiesbaden, Germany, 2018; pp. 897–920.
13. Bundesnetzagentur. Beendete Ausschreibungen. Available online: https://www.bundesnetzagentur.de/SharedDocs/Downloads/DE/Sachgebiete/Energie/Unternehmen_Institutionen/Ausschreibungen/Hintergrundpapiere/Statistik_Onshore.xlsx;jsessionid=CBEB8E4AF8685DF558604B2020D25E1F?__blob=publicationFile&v=1 (accessed on 14 June 2018).
14. Bundesnetzagentur. Pressemitteilun. Ergebnisse der Ersten Ausschreibung für Wind an Land. 19 May 2017. Available online: https://www.bundesnetzagentur.de/SharedDocs/Pressemitteilungen/DE/2017/19052017_Onshore.html (accessed on 14 June 2018).
15. Bundesnetzagentur. Pressemitteilung. Ergebnisse der Zweiten Ausschreibung für Wind an Land. Available online: https://www.bundesnetzagentur.de/SharedDocs/Pressemitteilungen/DE/2017/15082017_WindAnLand.html (accessed on 14 June 2018).
16. Erneuerbare Energien. Interview mit UKA-Chef. Ich Bin für Einen Funktionierenden Markt. 2017. Available online: https://www.erneuerbareenergien.de/ich-bin-fuer-einen-funktionierenden-markt/150/434/105190/6 (accessed on 23 April 2018).
17. Online Handelsregister (Trade Register). Available online: www.online-handelsregister.de (accessed on 23 April 2018).
18. Herrmann, N. *Heilt Die EEG-Notoperation den Fehlstart des Ausschreibungssystems?* EMW, Inc.: Herndon, VA, USA, 2017; Volume 4, pp. 1–5.
19. Die Welt. Die Schmutzigen Tricks mit der Bürgerenergie. 2017. Available online: https://www.welt.de/wirtschaft/article165807760/Die-schmutzige-Trickserei-mit-der-Buergerenergie.htm (accessed on 20 August 2018).

20. Deutscher Bundestag. *Antrag der Fraktion Bündnis 90/Die Grünen. Ausbau der Windenergie Sichern, Klimaschutz Voranbringen und Standort für Zukunftstechnologie Erhalten*; Drucksache 19/450. 17.01.2018; Deutscher Bundestag: Bonn, Germany, 2018.
21. Berkhout, V.; Cernusko, R.; Grashof, K. *Ein Systemwechsel ohne Vorteile*; Energie und Management: Herrsching, Germany, 2018; Available online: https://www.energie-und-management.de/nachrichten/alle/detail/ein-systemwechsel-ohne-vorteile-123389 (accessed on 14 June 2018).
22. Deutscher Bundestag. *Gesetzentwurf der Fraktionen der CDU/CSU und SPD*; Entwurf eines Gesetzes zur Einführung von Ausschreibungen für Strom aus Erneuerbaren Energien und zu Weiteren Änderungen des Rechts der Erneuerbaren Energien; Drucksache 18/8860, 21.06.2016; Deutscher Bundestag: Bonn, Germany, 2016.
23. Bundesverband Wind Energie. Ausschreibungen—EU-Wettbewerbskommissarin zum Thema de-Minimis. 2016. Available online: https://www.wind-energie.de/presse/meldungen/detail/ausschreibungen-eu-wettbewerbskommissarin-zum-thema-de-minimis/ (accessed on 15 January 2018).
24. Hauser, E.; Hildebrand, J.; Dröschel, B.; Klann, U.; Heib, S.; Grashof, K. *Nutzeneffekte von Bürgerenergie. Eine wissenschaftliche Qualifizierung und Quantifizierung der Nutzeneffekte der Bürgerenergie und ihrer Möglichen Bedeutung für die Energiewende*; Greenpeace Energy eG and Bündnis Bürgerenergie e.V.: Saarbrücken, Germany, 2015.
25. European Commission. *Proposal for a Directive of the European Parliament and of the Council on Common Rules for the Internal Market in Electricity*; 2016/0380 (COD), 23.2.2017 COM(2016) 864 final/2; European Commission: Brussels, Belgium, 2017.
26. Council of the European Union. *Proposal for a Directive of the European Parliament and of the Council on Common Rules for the Internal Market in Electricity*; Interinstitutional File: 2016/0380 (COD), 20 December 2017, 15886/17; European Union: Brussels, Belgium, 2017.
27. Wigand, F.; Förster, S.; Amazo, A.; Tiedemann, S. Auctions for Renewable Support: Lessons Learnt from International Experiences. Report D4.2, June 2016. Available online: https://www.ecofys.com/files/files/aures-wp4-synthesis-report-final.pdf (accessed on 20 May 2018).
28. Eberhard, A.; Kolker, J.; Leigland, J. South Africa's Renewable Energy IPP Procurement Program: Success Factors and Lessons. 2014. Available online: http://www.gsb.uct.ac.za/files/ppiafreport.pdf (accessed on 20 May 2018).

© 2018 by the author. Licensee MDPI, Basel, Switzerland. This article is an open access article distributed under the terms and conditions of the Creative Commons Attribution (CC BY) license (http://creativecommons.org/licenses/by/4.0/).

Article

# A Conceptual Framework for a Building Integrated Photovoltaics (BIPV) Educative-Communication Approach

Daniel Efurosibina Attoye [1,*], Timothy O. Adekunle [2], Kheira Anissa Tabet Aoul [1], Ahmed Hassan [1] and Samuel Osekafore Attoye [3]

1. Department of Architectural Engineering, United Arab Emirates University (UAEU), P.O. Box 15551, Al Ain 15258, UAE; kheira.anissa@uaeu.ac.ae (K.A.T.A.); ahmed.hassan@uaeu.ac.ae (A.H.)
2. Department of Architecture, University of Hartford, 200 Bloomfield Avenue, West Hartford, CT 06117, USA; adekunle@hartford.edu
3. Department of Mechanical Engineering, Indiana University-Purdue University Indianapolis, 317-274-5555 420 University Blvd, Indianapolis, IN 46202, USA; soattoye@iu.edu
* Correspondence: 201590088@uaeu.ac.ae; Tel.: +971-53-971-6472

Received: 22 August 2018; Accepted: 10 October 2018; Published: 19 October 2018

**Abstract:** Global interest in Building Integrated Photovoltaics (BIPV) has grown following forecasts of a compound annual growth rate of 18.7% and a total of 5.4 GW installed worldwide from 2013 to 2019. Although the BIPV technology has been in the public domain for the last three decades, its adoption has been hindered. Existing literature asserts that proper information and education at the proposal or early design stage is an important way of addressing adoption barriers. However, there is a lack of BIPV communication approaches for research, and market proposals that focus on clear information about its benefits. This has limited the adoption of BIPV.. Based on this, the present study aims to develop a conceptual framework for an educative-communication approach for presenting BIPV proposals to encourage its adoption. This is aimed at developing holistic research and market proposals which justify scholarly investigation and financial investment. Using a multiple case study investigation and Design Research Methodology (DRM) principles, the study developed an approach which combines core communication requirements, the pillars of sustainability and a hierarchical description of BIPV alongside its unique advantages. A two-step evaluation strategy involving an online pilot survey and a literature-based checklist, was used to validate the effectiveness of the developed approach. Our results show that understanding environmental and economic benefits was found to be significantly important to people who are likely adopters of BIPV ($p < 0.05$), making these benefits crucial drivers of adoption. Statistical significance was also found between those who do not know the benefits of using solar energy for electricity, and interest in knowing these benefits ($p < 0.05$). We thus conclude that proper communication of these benefits can safely be advanced as important facilitators of BIPV adoption. In general, this study elaborates the need and strategies for appropriate dissemination of innovative ideas to encourage and promote adoption of technological advancement for a sustainable global future.

**Keywords:** Building Integrated Photovoltaic (BIPV); barriers; sustainability; multi-functionality; proposal; educative-communication approach

## 1. Introduction

Innovation in the photovoltaic industry has spurred the growth of Building Integrated Photovoltaics (BIPV). BIPV refers to the use of photovoltaic (PV) devices to replace conventional building components of the building envelope, such as the roof, skylights or facades [1].

This technology brings a unique set of qualities and opportunities to the building industry such as on-site renewable energy generation, energy autonomy, and material multi-functionality. On-site power generation addresses the transmission and conversion losses of utility-scale photovoltaics with power generation close to the primary load [2–5]. BIPV also provides users with a degree of energy security and autonomy, and encourages reduced levels of energy consumption [6,7]. Conceptual and validated studies have verified BIPV energy output which reaches the Passivhaus threshold of 120 kWh/m$^2$, thus capable of enabling the Zero Energy Building (ZEB) target [8]. As a multi-functional building component, it also allows for daylighting and view; serving as safety glass, a shading device or privacy screen [9–11]. Each component of the building skin -roofing, walls, glazing, cladding, and fenestrations; as well as other external devices provide opportunities for integrating PV into the building in various applications [12–16].

In another light, BIPV technology represents the opportunity for a triple advantage in architectural design. It harnesses solar energy, addresses some limitations of utility-scale PV and converts the building from an energy consumer to energy producer as a multi-functional component. In harnessing solar energy, it utilizes renewable energy from the sun which provides more energy in one hour than the all the people on earth require for a whole year [17,18]. It also provides decentralized on-site energy right next to the point of use, thus reducing transmission and conversion losses, as well as ancillary costs associated with utility-scale PV [2–5,19–21]. Additionally, it serves as a multifunctional energy-producing building component used for roofing, cladding, glazing or shading [1,12,13].

The global BIPV market witnessed a 35% growth between 2014 and 2015 from an estimated 1.5 GW to 2.3 GW [22]. However, the global contribution of BIPV to the energy capacity added by Solar PV in 2016 was 1%—being about 3.4 GW of the total from Solar PV—about 303 GW [22–24]. Thus, though BIPV technology has multiple benefits and has been in public domain for the last three decades, its adoption rate in the built environment is limited. An overview of twelve studies on BIPV adoption [25–36] was carried out in a previous work [37]; from this, six major BIPV adoption barrier categories were identified. Most of these studies agree and specifically identify BIPV adoption barriers which relate to education/information [25–28,32], product, and project database [25–28], economy [26–30], industry [30–33], and management [25,29,30,33]. Among these, reported findings reflect that the impact of education linked with sufficient information and knowledge about BIPV is the most crucial [37]. Specifically, the review shows that, both within professional and public circles, there is a lack of sufficient knowledge on design, cost issues and multi-functional benefits from the environmental to economic to the social dimensions of BIPV adoption. Consequently, adoption of BIPV as a versatile renewable energy source is limited.

*1.1. Research Aim and Significance*

In light of the foregoing, this paper aims to fill this gap in the literature by developing a conceptual framework for an educative-communication approach to inform and facilitate adoption about BIPV. The duality and applicability of the intended approach may assist in strengthening its importance as an adoption driver by going beyond simply providing information (educative dimension) to communicating meaning (communication dimension). This stated significance of this study is driven by the need to increase BIPV adoption globally as a means towards harnessing its multifaceted benefits while reducing the negative impact of buildings on the built environment. This assertion is based on the fact that buildings are responsible for over 40% of the total annual global energy consumption, 10% of all CO2 emissions [38–40], 30–40% of greenhouse gas emissions, 30–40% of solid waste generated and 20% of all water consumption [41]. These figures show that the building industry is in need of a strategy which reduces its negative environmental impact. In line with this, the unique advantage of BIPV is that it harnesses renewable energy from the sun and also potentially stems the negative impact of buildings on a global scale as it converts the building from energy consumers into energy producers [42].

Several strategic approaches have been developed to communicate the importance of solar technology and/or BIPV adoption. However, this research seeks to advance and develop a communication approach which adopts key concepts from related approaches, expands the core requirements based on literature and develops an improved case-specific solution. Our approach is specifically oriented towards facilitating initial proposal presentations on BIPV adoption to justify market investments and research investigations. From a market point of view, the need to justify financial investment cannot be overstated as companies are always concerned about the market feasibility of a product. Similarly, research investigations need to address both a research gap and a research need. Thus, BIPV proposals in both fields (market and research) need to be sufficiently justified and communicated in a manner which explains and warrants the required time and financial investments.

*1.2. Research Design*

To develop the proposed conceptual framework, we proceeded by conceiving the investigation as the first stage of developing a grounded approach and thus focused on a deductive overview of the related literature; next, the development of the approach, and finally its evaluation. The literature overview was carried out on topics inherently related when considering BIPV technology, added to this section was an outline of BIPV educational barriers to set the stage for clarifying the research objective. The method section involved a combination of case study methodology with the Design Research Methodology (DRM) by Blessing and Chakrabarti [43]. This section also adapted principles from [44,45] to justify the procedure applied, and was framed to synthesize core requirements to aid the proposed design. Finally, we evaluated the developed approach in line with DRM principles. This was done using a pilot study on potential BIPV clients, and a checklist deduced from existing literature to verify if the approach meets both "market" and "research" expectations respectively. Figure 1 below shows the various stages of the research design with color codes for the respective steps; blue for the literature background, yellow for the method section and green for the evaluation section of the investigation.

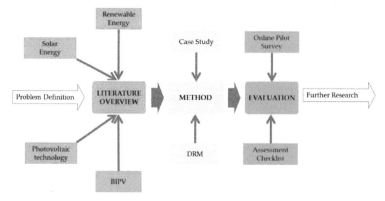

**Figure 1.** Research design.

## 2. Research Background

BIPV is an energy-producing building component which derives power from a renewable source (i.e., solar); it utilizes solar energy harnessed and converted into electricity by photovoltaic technology—via solar cells, which are integrated into the building envelope. At each level of mention as stated (i.e., renewable energy, solar energy, photovoltaics, and BIPV) significant research has been carried out in recent years. Energy production from Renewable Energy Systems (RES) is spreading but still represents a small part of the energy mix globally [46]. However, the simultaneous increase in energy demand and the negative impact of fossil fuels on the environment warrants the need for the development in this sector. A common thread between the existing studies on renewable energy, solar and photovoltaic technology is the identification of benefits, barriers, application, and future trends [47–50]. Research in these related areas suggests that key relationships can be traced and used to understand how the BIPV technology sits within the general body of research investigation relating to barriers and possible solutions.

Concerning renewable energy research, the authors of [47] assessed "limiting mechanisms" in this sector and identified the need for holistic cost comparisons between renewables and non-renewables as well as proper understanding of the ratio of energy produced to energy invested in forecast studies. The authors of [51] highlighted challenges and potential approaches towards the development of appropriate solutions for global renewable energy education initiatives. Among several findings, they noted a lack of structured curricula as a challenge, and thus suggested sufficient content in all renewable energy sources with local context emphasis as a desirable approach. About the attitudes of local communities towards investments in renewable energy projects, (i.e., adoption) a case study investigation showed that information and perceptions are behavioral determinants [46]. The authors asserted that a balance between economic, technical and environmental considerations in using RES would enhance more sustainable development towards future generations. To aid decision support for renewable technology and energy planning, the authors of [52] note the growing interest in flexible and user-friendly methods and a need for validation of results by "development of interactive decision support systems and application of fuzzy methods."

Other authors [48] highlight the fact that rapid growth in the field of solar technology, being a type of RES is marred by technical barriers such low solar cell efficiencies and performance of balance-of-systems (BOS) as well as economic and institutional obstacles. This persists even though some consider the sun as the most free and abundant renewable source [53] releasing $3.8 \times 10^{23}$ kW/s [54] and providing in an hour, more energy than the annual total global energy demand [17,18]. Some researchers [55–57] also note that there is limited awareness about the potential benefits of the solar industry in rural regions and that technology, research, and policy are identified as major challenges [49]. With various unique solar technologies such as solar PV, solar thermal, and solar fuels technologies as well as concentrated solar power (CSP), it has been suggested that choice of type, should be based on the type of usage/demand and prevailing conditions [58]. Also, technical parameters such as tilt and azimuth angles need to be better understood via design simulation [59]. The authors of [60] suggest that a unified approach to continued research, engineering and manufacturing will need to be pursued for the full potential of solar energy to be realized. In other terms, it has also been suggested that multidisciplinary approaches, perspectives, and collaborations are required to resolve these said barriers with solar technology applications [49].

On the other hand, research and developmental progress in solar power generation have been made in areas of hardware development and testing towards efficiency maximization and cost minimization [50]. Challenges such as affordability, needed policies, appropriate system planning [50,61] have warranted suggestions and survival strategies for PV technology [62] such as new developments, improvements and innovations [63,64]. In order to improve PV self-consumption, the authors of [65] noted that energy storage and load management, also called demand-side management (DSM) are needed considerations. To increase the participation of photovoltaic energy in the renewable energy, raising market awareness to its benefit has also been suggested [61].

The limitations mentioned above are inherent in related topics of interest when considering BIPV technology and are crucial. These topics form its intrinsic constitution and thus require active research and development to address them. The referenced studies suggest several crucial points, one being that there is a need for sufficient, credible and contextual information to be presented to stakeholders to aid adoption of the technology. This should also be synthesized and presented using flexible and user-friendly methods which embrace a multidisciplinary approach drawing from the environmental, economic and social dimensions. Besides the above, however, the BIPV technology also has its unique barriers and limiting factors as identified in several studies mentioned in the introduction. These barriers are in areas of education, product, economy, database, industry, and management with previous research identifying BIPV education/information awareness as the most crucial of these [37]. Seeing that the interposing barriers mentioned above (i.e., in line with RES, solar and PV) also identify information, awareness, and education as a need, this study seeks to address this challenge by the development of the proposed Educative-Communication Approach. The educative dimension provides contextual information, and the communication dimension is meant to facilitate interaction to inspire BIPV adoption.

*Education and Communication of BIPV as a Technological Innovation*

With the specific mention of BIPV educational barriers mentioned above, literature shows that both professional and public domains are affected. Specifically, findings from surveys on public educational barriers state various reasons such as a poor public understanding of cost perceptions of BIPV and financial benefit understanding [29,33] and a lack of sufficient knowledge by clients and the public in general [26,33]. Also reported is a high negative perception of system price and costs associated with aesthetic BIPV options [33]. In 2017, a survey was conducted in Europe to identify educational needs to resolve this barrier [27]. The report identifies a dearth of university courses on BIPV and also identifies the lack, and thus the need for knowledge of product options, design strategies, performance and cost issues. Also refs. [26,28,30] respectively report that a lack of sufficient technical knowledge, certified BIPV contractors available, and insufficient knowledge about BIPV system advantages, risk, and complexity exists. The lack of knowledge on how to ensure the most efficient choice of BIPV design has also been noted [31].

Having stated the above, providing education and information to aid BIPV proposals—market or research requires proper communication. It has been reported that "the successful design and realization of solar architecture—in general, relies upon the effective communication of its qualities in the development of a project" [35]. Solar energy systems are becoming more cost-efficient through continuing technological development [50,62]. However, the growth of these systems into a comprehensive and everyday solution, to the zenith of being a "natural choice for all projects" is consequent on the proper communication of the qualities of these systems [35]. Mastering the best balance between installed power, energy generation and aesthetic appearance of solar technology is not an easy task, and the lack of information will be decisive [66]. Communicating goals, information and idea sharing with stakeholders has been advocated as a means to facilitate a sufficient and mutual understanding. Indeed, it has been asserted that project goals, from inception to completion are facilitated via proper communication [35]. Given its multiple advantages earlier stated, planned and evaluated investment of BIPV, as well as the adoption of convincing approaches should therefore be encouraged. This will be needed to justify solar energy adoption in project proposals, especially as the energy solution and the use of solar energy are a significant part of the considerations when designing a building.

At present, the BIPV technological innovation is in a cycle of debate and development, which is, indeed, characteristic of all innovative ideas. This investigation presents the relevance of both educative/informative and communicative goals towards facilitating adoption as its theoretical underpinning based on referenced related literature. The "diffusion of innovation" theory by Everett Rogers has been reviewed [67] and it asserts that relative advantage (i.e., benefits) is the strongest predictor of the rate of adoption of an innovation. The term 'Relative Advantage' suggests the degree of perceived advantage or state of being better than existing options [68]. The authors of [69] carried out a detailed review of 20 innovation frameworks and conclude that determinants which potentially foster adoption include clarity and simplicity of use as well as improved benefit over existing options. They also assert that clear research evidence, cost-efficacy and feasibility, relating to expected benefits are also crucial drivers which advance the adoption of innovation. These theorists thus agree that educating and communicating benefits is a multi-disciplinary philosophy which is a potential driver of innovation adoption.

## 3. Method

As earlier stated, the method of investigation combines a case study and the DRM principles to carry out the design and development of the proposed approach, followed by an evaluation and review. This 3-phase approach is detailed in this section; representing the foundational and conceptual design process for the proposed approach.

### 3.1. Phase 1: Case Study Investigation on BIPV Communication Approaches

Following the categorization of case study typologies [44], we carried out a descriptive qualitative overview of multiple case studies. Several studies referenced below distinguish between single and multiple case studies. The consensus is that multiple-case studies generally enhance a stronger base for theory building [70], facilitate comparisons to elucidate whether an emergent finding is case-specific or subject-specific, i.e., replicated by several cases [71]. Other advantages include wider elaboration and exploration of research questions and theory [45]. General case study selection considerations from literature which guided our investigation include:

- Ability to reflect characteristics identified in the underlying theoretical background/propositions [44]
- Ability of the chosen cases to fit the purpose of the research [72]
- Ability to represent and exemplify the phenomenon of inquiry [72]
- Suitability for illuminating and extending relationships related to the investigation [45]
- Applicable for detailing and expanding logic among constructs of the study [45]
- Clear description of the existence of a phenomenon [73]

To address the theoretical sampling complexity inherent in multiple cases studies, existing literature suggests that the choice is based less on case uniqueness but more on the contributory relevance, agreeing with theoretical propositions and development within the set of cases [44,45]. Based on these requirements, we reviewed related studies on BIPV adoption, and we identified three (3) studies which focus on developed approaches to communicate BIPV. These include the International Energy Agency—IEA Task 41-SubTask C three-stage approach from client to design team to design-communication tools [74], the European-based use of "an ambitious demonstration project portfolio" [75], and an Architectural Integration Qualities (AIQ) model which initiates and focuses discussions on preferences for architectural integration of energy-producing solar shading [76]. The sub-sections which follow state an overview of these studies used to deduce core requirements for the proposed approach as a starting point for the design process.

### 3.1.1. The IEA TASK 41 Communication Guide

IEA Task 41, "Solar Energy and Architecture" focused its subtask C on developing communication guidelines and delivered a 3-step approach for achieving this goal. These steps, laid out as sections in the Guide [74], can occur continuously during design development (Figure 2). Indeed, the report posits that one of the main reasons for not adopting solar energy in projects today is the lack of client confidence in the field. It was suggested that investors do not generally adopt solar energy in projects due to lack of knowledge, experience, and accessible information about benefits, risks, and system characteristics.

The first section of the guideline focuses on strategies to convince clients to request solar building projects. It includes recommendations for identifying client goals and motivations, and important integration considerations for common project types. Section two addresses communication strategies at the design and construction team level. It suggests techniques for "anchoring solar energy strategies within the project team" and communication strategies with manufacturers, with added content for strategies within a design-build process. Section three discusses tools for communication and design development with included national references to design guidelines for solar energy and architecture within the Task's participating countries. This case study provides a holistic, macro-level consideration of the BIPV communication approach.

Figure 2. International Energy Agency (IEA) Task 41 three-step approach. Source: Adapted from Ref. [74].

3.1.2. The PVSITES Project

The PVSITES project began in 2016 as a joint multi-disciplinary European approach to drive large market deployment of BIPV technology. The goal is to demonstrate "an ambitious portfolio of building-integrated solar technologies and systems." This was towards giving a definite and professionally reliable answer to the market requirements pre-identified by the industrial members of the consortium. It focuses on high impact demonstration and dissemination actions to be executed using cost-effective renewable generation, reduction of energy demands and smart energy management [75,77].

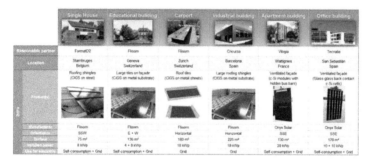

Plate 1. Overview of PVSITES' demonstration projects. Source: [75].

The authors of [75] argue that the gradual acceptance of BIPV installations by stakeholders requires the use of high visibility showcase projects. In the PVSITES demo cases, the proposed module, grid interface, and energy management technologies are demonstrated in at least six real buildings covering different EU electricity markets and climates, building uses (residential, industrial, commercial) in new buildings and retrofit projects. These are designed to highlight different architectural integration strategies, allow for easy replication and adaptation. The buildings showcased in Plate 1 above, show a varied range of innovative BIPV products—roof shingles, roof and façade tiles, and a ventilated façade—in different European locations such as Belgium, Switzerland, Spain, and France [65]. This case study provides a design-focused, macro-level consideration of the BIPV communication approach facilitated by completed projects.

3.1.3. The AIQ-Model

This case study focuses on the development a tool to aid communication on the subject of "energy-producing solar shading" as this relates to a type of BIPV (i.e., BIPV shading devices) [76]. It is based on the findings of other authors who argue that BIPV design is in need of a multi-disciplinary communication language or tool [10,64]. Therefore, an approach to evaluate architectural integration qualities was developed to communicate the aesthetic values of such systems. The AIQ-Tool developed is visualized as a triangle where corners represent geometry, materiality and detailing; to evaluate the degree (poor, fair or good) to which solar shading system fits with the overall building design.

The authors report that the AIQ-Tool helped the surveyed participants of a test focus group to articulate architectural integration and gave rise to positive discussions. Also, it was observed that although architects had the best conditions to use the tool, the discussions were facilitated by other professions. A critique of the tool as presented by its developers is its limitation to "external integration and aesthetics". They state that this does not embrace the inter-disciplinary perspectives of solar building design which requires multi-functional energy planning as perceived by different disciplines. This case study asserts the need for an expanded consideration across disciplines to provide an inclusive and holistic BIPV communication approach.

3.1.4. Case Study Deductions

To extract deductions from these cases and achieve some preliminary analysis, we aimed at a synthesis of the characteristics based on analytical generalization following "replication logic". This was done in agreement with literature recommendations "to generalize theoretical propositions and not a population as in statistical research [44]. Apart from guiding an identification of core requirements for the proposed approach, the synthesis was also to provide suggestions and show limitations to avoid in the design of the approach. The following list deduced represents applicable key points, used in the reference cases to boost proper communication before, during or after the design phase.

1. Provide accessible information about BIPV benefits and risks [74,75]
2. Convince clients to request BIPV [74,75]
3. Anchor solar energy strategies within the project [74]
4. Maximize tools or models for communication [74,76]
5. Apply a continuous communication process [74]
6. Utilize high impact demonstration projects [75]
7. Apply a multi-disciplinary communication tool [75,76]

These seven points are used as a guide to the design and as part of the strategy to review the developed educative-communication approach in the evaluation section. This was done in addition to other associated requirements for addressing BIPV barriers in tandem with its RES-Solar-PV roots extracted from the literature section. The combined requirements were distilled into an evaluation checklist in the final section of this investigation to assess the effectiveness of the developed approach.

*3.2. Phase 2: Design of the Approach*

The DRM approach by [43] is a multi-disciplinary guide which provides a framework for design research, development of the research argumentation, guidelines for research planning and methods. Within the approach, the "Development of Support" method (Type 3) is recommended for use when the existing literature review provides sufficient guide to start the design development. As this scenario matches our present investigation, we adopted this 'research type'. The term "support" in this case refers to an aid used to improve design such as strategies or approaches and the process for introducing methods. In our research scenario, existing support options for BIPV communication to improve chances of adoption are present but limited and case-specific proposal presentations are not prioritized. Therefore, we opted to integrate the principles of the systematic Prescriptive Study (PS) which is recommended for use when the existing support is insufficient.

Our adaptation of the PS is justified by the flexible and adaptable nature of the DRM principles to match project requirements, and covers three core areas of task clarification, conceptualization, and preliminary evaluation. Our introduction and literature overview provided an awareness of the problem statement to clarify the design task (i.e., task clarification). For the conceptualization stage of the conceptual framework intended, the case study in Section 3.1 provided the core requirements for the support. We thus, required a firm concept, based on the DRM system to frame the design of the proposed approach. The DRM system allows for assumptions and experience to guide this selection which in generic terms ought to satisfy applicability, adaptability and generalizability to the task under review. In the current era of sustainable development, decision making in energy projects requires consideration of technical, economic, environmental and social impacts [51,78–81]. The process of energy planning under this largely interdisciplinary scenario is thus, often complicated due to the involvement of these multiple benchmarks [82]. In line with these considerations, we required a broad-based multi-disciplinary concept. We thus, chose the pillars of sustainability and a redefinition of the BIPV structure as our theoretical driving concept. The following section clearly explains how this informs the process of designing our conceptual framework.

3.2.1. Step 1: Pillars of Sustainability as the 1st Driving Concept

The mainstream theory for sustainability has become the idea of three pillars (3Ps) namely: economic, social and environmental sustainability [83]. Although there are other cultural and political aspects, these do not form the core of our focus in this investigation. The pillars of sustainability follow the concept that every sustainable approach or idea must provide benefits regarding the cost, social impact, and ecological impact or carbon footprint. The three pillars are interwoven and have been explained in different ways to highlight the importance of sustainability and the three major players -people, planet, and profits. Adopted by the General Assembly of the 2002 and 2005 World Summit on Sustainable Development, these three components—economic development, social development, and environmental protection—are presented as interdependent and mutually reinforcing pillars [84,85]. Today, these pillars are expressed and discussed extensively across various governmental, professional and commercial circles; influencing concepts like the triple bottom line in sustainable urbanism and other aspects of the sustainable built environment. They respectively relate to continued support for a defined economic production, maintenance of social well-being, and ability to ensure responsible use of renewable resources to curb non-renewable resource depletion.

A few extra points to note; environmental sustainability suggests that the framework must promote the overall well-being of people. For the social sustainability, the concept must maintain equity while economic sustainability ensures the framework is not only innovative but cost-effective. Based on the definitions of the pillars, it is essential to state that any connecting framework or approach must meet the requirements highlighted in the descriptions. As such, our investigation agrees that for BIPV, the framework for research and market proposals must satisfy the crucial requirements for the pillars of sustainability. Also, the integration of the pillars for the development of a framework must provide a truly sustainable design or development that will make the world a better place.

3.2.2. Step 2: BIPV Triple Advantage and Hierarchy of Form as the 2nd Driving Concept

A detailed understanding of BIPV viz-a-viz, a structural breakdown of its constituents, has been suggested [86]. Reference is made to the elemental and compositional dimensions; the former relates to specifics such as the cell technology, cells shape, module design, and arrangement. The latter refers to the building function and type of product. In addition to this descriptive Index, a holistic understanding of BIPV can encompass the hierarchy of BIPV origins and form. The hierarchical composition of BIPV earlier mentioned in the introduction refers to it as a building component, next as type of PV technology; then as a strategy which harnesses solar energy to generate electricity. Further, solar energy is itself a renewable source of energy which assists to reduce the use of non-renewables and stem the rate of global environmental pollution. This breakdown forms the philosophical idea put forward as a part of the theoretical concept for this investigation. It portrays a broader perspective of what BIPV represents and may help to appreciate its relevance to society and facilitate its adoption. The ability to communicate relevance and importance within a proposal context, at each aspect of this chain represents a holistic understanding of the importance of BIPV. Figure 3 shows a diagrammatic illustration of the BIPV-PV-Solar-Renewable chain.

**Figure 3.** Building Integrated Photovoltaics (BIPV) Hierarchy of Form. Source: Authors.

### 3.2.3. Step 3: The BIPV-3P Matrix

For the final stage of the "Approach development" we designed an integrated matrix which presents a simple juxtaposition of the BIPV hierarchy of form with the 3 Pillars of sustainability. This leads to a comparison of the four components of the BIPV-PV-Solar-Renewable chain with the Environmental–Economic–Social Pillars. In this comparison, the BIPV technology/proposal/project/is discussed at each level of its hierarchy based on associated environmental, social or economic benefits. Added to the approach is the design dimension to simulate the intrinsic architectural orientation of BIPV. Figure 4 below shows the diagrammatic color-coded representation of the matrix. Each cell in the matrix corresponds to the required information at each level of the BIPV Hierarchy based on its 3P benefits. The grid format selected assists in a structured and systematic approach to present the facts required to justify the project/proposal objectives and benefits.

The matrix brings together all the work done so far in a singular figurative depiction and forms the illustrative representation of the conceptual framework. Its visual depiction can be modified based on the objectives of the proposal in view to match the kind of audience being discussed with. However, its objective and intent remain holistic and inclusive of the focused findings from literature which relate to specifics along the BIPV-Solar-PV-Renewable chain. It also integrates the core requirements deduced from the case study and follows the systematic approach recommended by the DRM principles applied.

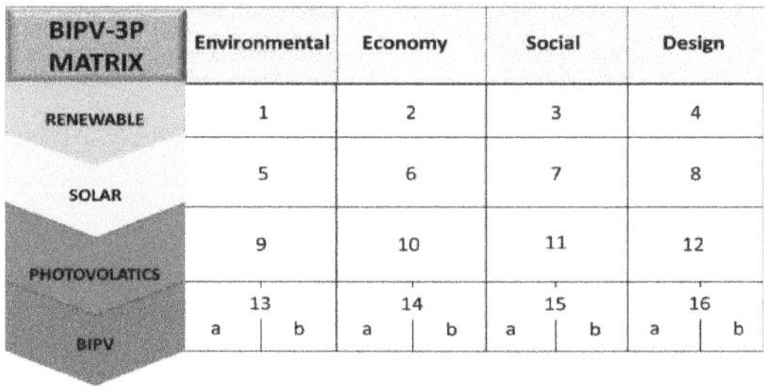

| BIPV-3P MATRIX | Environmental | Economy | Social | Design |
|---|---|---|---|---|
| RENEWABLE | 1 | 2 | 3 | 4 |
| SOLAR | 5 | 6 | 7 | 8 |
| PHOTOVOLATICS | 9 | 10 | 11 | 12 |
| BIPV | 13 a \| b | 14 a \| b | 15 a \| b | 16 a \| b |

**Figure 4.** BIPV-3P Matrix. Source: Authors.

### 3.3. Phase 3: Evaluation

As earlier stated, in agreement with standard project planning and the DRM principles, we carried put an evaluation to review and if needed, improve the developed conceptual framework. To achieve this preliminary evaluation; two steps were taken; one to check its effectiveness towards actually facilitating the adoption of BIPV, the other to check its agreement with recommendations from literature. The first step was an online pilot survey redefined as a prospective user experience (UX) survey; the second, the use of a "Case Study Deductions Checklist" developed from the literature section. Both strategies were conceived and carried out as a part of this research, and the results are summarized below.

3.3.1. Pilot Survey: User Experience (UX) Format

User Experience (UX) is defined as "a person's perceptions and responses that result from the use or anticipated use of a product, system or service" [87]. The authors of [88] also describe UX as "a consequence of a user's internal state (e.g., predispositions and expectations), the characteristics of the designed system (e.g., complexity, usability) and the context within which the interaction occurs". UX research thus allows investigators to carry out qualitative research while studying user behavior–actual or anticipated. Following our earlier description, one frame of reference (i.e., driving concept) used in this investigation to develop the approach is the "pillars of sustainability", i.e., 3P. This study anticipated a UX survey as a part of its evaluation of the conceptual framework while asserting a scholastic argument. This argument, advanced in this study is that persons who have an understanding of these 3P benefits/dimensions of BIPV will likely be positively disposed towards its adoption. This assumption agrees with several studies [25–27] and is also based on the innovations theories earlier discussed. Following this, we considered it logical to facilitate a means of engaging public feedback to evaluate the developed approach. To achieve this, a pilot UX-type survey was run on the internet for one week (10–17 June 2018). It was randomly deployed via social media—specifically Facebook and WhatsApp, with the questions prepared using Google Forms. Sixty-nine (69) responses were received, none were invalid, and all were used in the analysis. The objective of the survey was to confirm if the developed approach was indeed able to encourage BIPV adoption. It was thus designed to find out public opinion about electricity for home/office/school using solar panels as a part of the building. In statistical considerations, we also sought to investigate if there is a significant relationship between knowledge of BIPV benefits and the decision to adopt BIPV to justify the theoretical position of this investigation.

For clarity, the questions were designed to ensure proper understanding by both professionals and non-professionals. As such, technical details were redefined in simple everyday terms as explained below. The 3P concepts towards BIPV benefits i.e., environmental, economic and social were simplified; "environmental benefits" was simply framed as "helping the environment", "economic benefits" as "saving money", and "social benefits" as "higher social status and recognition". However, to ensure a holistic representation of these benefits, and guarantee that related aspects were not omitted, respondents were allowed to state any benefit they also considered important to adopting BIPV. In general, the underlying assumption of the simulated UX survey is that if a respondent considered that a benefit is important, then it has a potential impact on their decision to adopt BIPV. The list that follows is the results of the survey for each of the three questions, followed by a brief analysis of the results.

1. Question 1: Do you know the benefits of using solar energy for electricity?

The purpose of this question was to introduce and affirm our subject to the respondent, as well as establish a basis for further inquiry. It was framed to determine if the respondent was aware of the concept and availability of solar energy as a renewable energy system. A "yes" or "no" answer was provided to the respondents. The survey did not consider an "indifferent" or "no answer" option regarding this question as knowledge, not opinion was required at this point. From the results, over 95% (66 respondents) chose the option showing they had an understanding of the benefits of solar energy as a source of power. Only about 4% selected the option stating they did not know the merits of solar energy. This suggests a high level of awareness of the sustainable characteristic of solar energy amongst the respondents. While the 95% represents the group of persons who are considered "likely to adopt BIPV" based on the knowledge about its benefits, 4% are "unlikely to adopt". This characteristic of these groups makes for significant importance to this study and is later discussed further.

2. Question 2: Which of these benefits is most important to you when deciding to use solar energy for electricity?

Respondents were asked to present their perceived importance of solar energy benefits. A multiple-choice format presented possible benefits relating to the 3P concept used to develop the approach. These were, helping the environment, better building design, saving money, higher social status and recognition. Respondents were permitted to select more than one option or add to the list.

Figure 5 above shows that over 81% (54 respondents) felt that solar energy was beneficial for two main reasons; firstly, by using it, they are helping the environment and secondly saving money. 24% (16 respondents) noted that better building design was a desirable benefit. Only about 5% felt all benefits were significant, and only about 2% (1 respondent) felt a higher social status and recognition was an important benefit. Six respondents (9%) mentioned other specific reasons they would consider, but on closer examination, they mainly relate to design, environmental, and social issues already listed.

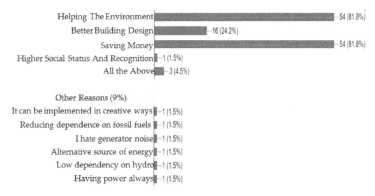

**Figure 5.** Responses to Question 2.

3. Question 3: Will you like to know these benefits?

This question was specifically reserved for respondents who answered: "NO" to question 1 ("do you know the benefits of using solar energy for electricity"?). This question was to identify if they desired to increase their knowledge of the merits of solar energy. All the three respondents (100%) who answered "NO" in question 1 answered in the affirmative to this question (i.e., they will like to know these benefits).

Further statistical analysis of the pilot—shown in Table 1 below—revealed the mean values of Questions 1 to 3. The table shows the standard deviation values, sample variances, level of confidence for the variables. A significance was found between the respondents that did not know the benefits of using solar energy for electricity and would like to know the benefits ($p < 0.05$). Regarding Question 3, the statistical analysis considered the responses; the results showed the average respondents considered helping the environment and saving money as the most crucial benefits when deciding the to use the solar energy for electricity. A relationship was therefore found between the respondents that understand the benefits of using solar energy for electricity and the respondents that selected helping the environment and saving money ($p < 0.05$). The import of this relationship shows that understanding environmental and economic benefits was found to be significantly important to people who are likely adopters of BIPV, making these benefits crucial drivers of adoption.

Table 1. Mean, standard deviation, variable, and level of confidence for the pilot survey.

| Questions/Variables | Number of Respondents | Mean | Standard Deviation | Sample Variance | Level of Confidence at 95% |
|---|---|---|---|---|---|
| Q1: Do you know the benefits of using solar energy for electricity? Note: 1—Yes, 2—No | 69 | 1.0435 | 0.02473 | 0.0422 | $p < 0.05$ |
| Q2: Will you like to know these benefits? Note: 1—Yes; 2—No | 69 | 1.9565 | 0.02054 | 0.04219 | $p < 0.05$ |
| Q3: Which of these benefits is most important to you when deciding to use solar energy for electricity? (You can select more than one) | 69 | 7.4559 | 0.4377 | 13.0279 | The overall $p$-value is greater than 0.05 |

Note: Q3: 1–Helping the environment; 2–Better building design; 3–Saving money; 4–Higher social status and values; 5–All of the above; 6–Other reasons; 7–Helping the environment; Saving money; 8–Better building design; Saving money; 9–Helping the environment; Better building design; Saving money; 10–Helping the environment; Better building design; Other reasons; 11–Saving money; Other reasons; 12–Helping the environment; Other reasons; 13–Helping the environment; Better building design; 14–Helping the environment; Saving money; All of the above; 15–Helping the environment; Saving money; Other reasons

The results of the pilot survey show that most of the respondents know the benefits of using solar energy for electricity as part of the building. Also, environmental and economic benefits are considered most significant and social benefits less important by respondents interested in adopting BIPV. For the few who do not know the benefits of using solar for electricity, they were all interested in learning. Based on these findings, we tentatively conclude that since the approach provides copious information about environmental and economic benefits, it will potentially encourage adoption—at least for the respondents of the survey.

3.3.2. Case Study Deductions Checklist

The second evaluation strategy was a checklist developed to review the features of the BIPV communication approach developed based on the core requirements deduced from the case study section. This checklist also combines deductions and extrapolations from existing literature which describe requirements for proper BIPV communication. For each item on the Checklist, a "Yes" or "No" is given based on the structure and contents of the developed approach (See Table 2). A remarks column is also provided to give extra information as it relates to each specific requirement.

This checklist does not intend to cover all the aspects and requirements for BIPV communication. However, it satisfies certain crucial points which have been satisfied by the approach developed. These include the following:

- It helps to present the important features of the developed approach
- It serves as a means of evaluating the effectiveness of the approach and any other of the same basic requirement, format or approach
- It establishes an agreement between requirements in literature deduced from established projects and the features of the approach
- It confirms that the goal of the investigation to facilitate case-specific and contextual communication on BIPV has been achieved.
- It proves that the approach is research-based and thus a credible means of communicating benefits of a BIPV proposal.

**Table 2.** Checklist for a BIPV communication approach. Source: Authors.

| | Checklist | Yes | No | Remark |
|---|---|---|---|---|
| 1. | Provide accessible information about BIPV benefits and risks [74,75] | ✓ | | The approach can be used to highlight BIPV benefits as well as benefits of RES, solar and PV. The information provided can also be used to assess and compare with other sources of energy to develop a comparative or risk assessment plan. |
| 2. | Convince clients to request BIPV [74,75] | ✓ | | The approach is based on a theoretical background which advances communication as a means to advance adoption. It also provides information which can be used to inform and educate clients on the merits of BIPV adoption. |
| 3. | Anchor solar energy strategies within the project [74] | ✓ | | To enhance understanding and representation of solar energy strategies, the approach anchors BIPV within a specific context for each proposal. It starts out general but ends with very explicit information on the project proposal. |
| 4. | Maximize tools or models for communication [74,76] | ✓ | | With a simple tabulated interface, the approach drives communication which is guided by a sequential presentation of relevant facts. Adapting this format to a digital interface is the next phase of its implementation. |
| 5. | Apply a continuous communication process [74] | ✓ | | By addressing multiple concerns; environmental, economic, social and design, the approach covers the various aspects of building design from conceptualization to completion. The matrix format may also guide review during iterative changes in the proposal as it shows hierarchy and relationship between columns and rows |
| 6. | Utilize high impact demonstration projects [75] | | ✗ | Although the approach can be applied to a wide range of projects including small, medium and large-scale; it does not directly represent a demonstration project. |
| 7. | Apply a multi-disciplinary communication tool [75,76] | ✓ | | Drawing from the multi-disciplinary nature of the BIPV technology, the approach attempts to show aspects and concerns of interacting disciplines. This can aid team building and brainstorming, and validate the need for each professional within the team. |
| 8. | Ratio of Energy [47] | ✓ | | The approach can assist to present a relative comparison between energy generated from BIPV and other energy sources to aid decision making and investment |
| 9. | Local context emphasis [50,75] | ✓ | | The primary focus of the cells relating to the BIPV system is specifically and locally contextualized to the project under review. The information can be used to compare various projects and inform management decisions |
| 10. | Balance between economic, technical and environmental considerations [45] | ✓ | | By using the pillars of sustainability as a driving concept, the approach presents a balance between these interrelated aspects and suggests that extended input can be made to show specific data within each of these considerations. |
| 11. | Flexible and User-Friendly Methods [51,76] | ✓ | | The simple matrix format assists non-professional to understand technical information without complex presentation. It also clearly utilizes a symbolic vertical and horizontal format to suggest the relationship between these related issues to facilitate planning. |

## 4. Discussion

The developed matrix is divided into a set of rows and columns to communicate the proposal/project idea. The information contained by detailing the 3P section of BIPV hierarchy 1 to 3 (i.e., Renewable, Solar and Photovoltaic aspects) is fundamentally similar for all projects (Cells 1 to 12). However, discussing it contextually can not only be different, but also potentially presents better relevance and aids understanding. For example, Renewable Energy (BIPV Hierarchy 1) has economic benefits (Cell 2) which are ultimately based on regional policies. As such, this information will differ for projects in separate geographical locations and consequently impact the contents of the matrix. BIPV Hierarchy 4 (Cell 13–16) is the core of the proposal, and the 3P outline should be discussed at two levels; firstly, the benefits of BIPV as an energy source and secondly, as a building component. To provide a better understanding of the matrix, crucial information required for communicating a BIPV project proposal for the sixteen cells has been outlined below (See Table 3). The list of suggested contents is aligned with the 3P columns on the matrix. However, this guide is not exhaustive, but rather, a list to showcase and justify the objective of the approach. Similar questions should be added to facilitate contextual and holistic potentials based on unique characteristics of individual proposals. For each of these cells, information is to be provided which is specific to the project proposal, with the background facts on the 3P benefits of renewable, solar and PV hierarchies. This matrix is flexible and can be presented as it is, or modified based on the specifics of the proposal. Although all the cells need not be filled, a general introduction of the BIPV hierarchy following the suggested chain can assist to develop a strong presentation to justify market/financial investments and research investigation.

Table 3. Suggested contents of the BIPV-3P Matrix. Source: Authors.

| Cells | Suggested Content |
|---|---|
| **Cells 1 to 12** | **BIPV Hierarchy 1 to 3** |
| Cell 1: Environmental benefits of Renewables | - State cumulative percentage/amount in tons of reduction of carbon emissions in the region<br>- State accrued benefits in wildlife conservation and human preservation (or related interest to sponsor) |
| Cell 2: Economic benefits of Renewables | - State fuel and maintenance cost savings compared with non-renewable energy sources<br>- State marketability of free natural resources |
| Cell 3: Social benefits of Renewables | - State the potential reduction in the Social Cost of Carbon (SCC) associated with similar energy output from a fossil fuel power plant<br>- State accrued benefits of replacing fossil energy sources, and other points such as international recognition and accountability |
| Cell 4: Design benefits of Renewables | - Highlight adopting buildings as a free-standing support medium for Building Integrated Renewables<br>- State potential visual impact on energy awareness on the residents in the region |
| Cell 5: Environmental benefits of Solar Energy | - State cumulative percentage/amount in tons of reduction in carbon emissions in the region<br>- State reduction in pollution (e.g., noise) during use compared to fossil fuel energy generation |
| Cell 6: Economic benefits of Solar Energy | - State energy security benefits and autonomy; and advantages of a constant source of fuel<br>- State flexibility and adaptability for basic household use and advanced technological applications |
| Cell 7: Social benefits of Solar Energy | - State potential to advance global energy reduction targets and advocacy/image recognition<br>- State potential for labor employment and other corporate social responsibilities |
| Cell 8: Design benefits of Solar Energy | - State passive opportunities such as daylighting, along with sustainability benefits<br>- State active opportunities such as photovoltaics, along with sustainability innovations |
| Cell 9: Environmental benefits of Photovoltaics | - State cumulative percentage/amount in tons of reduction of carbon emissions in the region<br>- State advantages of a constant source of fuel relating to the reduced recurrent need for fuel harvesting |
| Cell 10: Economic benefits of Photovoltaics | - State comparative long-term cost benefits compared with other energy sources relating to maintenance<br>- State savings in cost of fuel compared to other energy sources |
| Cell 11: Social benefits of Photovoltaics | - State investment as a form of social responsibility towards a global sustainable future<br>- State labor employment, advocacy, and support for the industry |
| Cell 12: Design benefits of Photovoltaics | - State opportunities as a building integrated or building applied system<br>- State technological growth as a sign of the global shift towards harmony with the architectural design and opportunity for clean energy from buildings |
| **Cell 13a–16a** | **BIPV Hierarchy 4: as an Energy Source** |
| Cell 13a: Environmental Benefits | - State how much the proposal reduces $CO_2$ emission<br>- State how much land is saved compared to utility-scale PV based on expected power output<br>- State the number of trees saved by using BIPV in the project based on similar expected power output from a utility-scale PV plant |
| Cell 14a: Economic Benefits | - State the amount of savings in labor cost<br>- State the amount of savings in infrastructure cost compared to utility-scale PV based on expected power output<br>- State the cost savings in land purchase compared to a utility-scale project of the same expected power output |
| Cell 15a: Social Benefits | - State the visibility of the project to the public<br>- State opportunities for educating the public<br>- State potentials to achieve significant recognition by prioritizing on regional /international sustainability ratings such as LEED (Leadership in Energy and Environmental Design) |
| Cell 16a: Design Benefits | - State the amount of energy produced<br>- State the amount of energy saved compared to use of non-renewable sources<br>- State the benefits of energy control enjoyed by the intended owners |
| **Cell 13b–16b** | **BIPV Hierarchy 4: as a Building Component** |
| Cell 13b: Environmental Benefits | - State the savings in embodied energy<br>- State the environmental impact advantage compared with replaced building materials<br>- State benefits as a type of quasi-modular construction system |
| Cell 14b: Economic Benefits | - State labor and other aggregated cost savings compared with alternative materials, e.g., bricks or blockwork; mortar, painting; and separate costs for glazing and associated costs.<br>- At an advanced level, carry out a full comparative life cycle analysis with other material alternatives<br>- State benefit for government payback if Feed-In-Tariffs (FITs) are regionally available |
| Cell 15b: Social Benefits | - State potential visual impact and energy awareness education on/for the residents in the region<br>- State potential to serve as contemporary green building icon<br>- State other potentials for household energy autonomy |
| Cell 16b: Design Benefits | - Discuss the aesthetic potential of the project compared with other surrounding modern buildings<br>- State potential for multiple integration opportunities on roof or facade<br>- State multi-functional uses of the BIPV installation: does it provide daylighting or view or shading along with energy for example. |

It is necessary at this stage to revisit the initial scenario depicted by existing literature and the potential relevance of this investigation. Information and perception have been listed as determinants in the adoption renewable energy project [45] and the balance of economic, technical and environmental benefits is crucial. The matrix represents a simple but clear approach to addressing these issues. It also helps in decision support for renewable technology and energy planning, built on and applying a user-friendly method [51]. The identification of limited awareness of solar energy benefits specifically in rural regions [56] can be addressed by applying the matrix as a simple-to-understand communication tool. It also embraces suggested multidisciplinary approaches and perspectives [48] while raising market awareness to its benefits [60]. The results of the pilot UX survey also agree with innovation theorists [68,69], and suggests that knowledge of benefits can promote the rate of adoption of an innovation among likely adopters. The priority given to environmental and economic benefits, over social and design benefits is subjective and is influenced by the perspective of the respondents. Further analysis of respondent or stakeholder type is required to fully understand these results. Nevertheless, the results are in general agreement with other studies on the perceived values attached to BIPV as an environmentally-friendly energy source with potential cost-saving benefits [1,29]. Conclusively, the conceptual approach and the matrix developed, addresses confidence issues in BIPV proposals by providing information, knowledge and access to verifiable information about system benefits.

## 5. Future Research

Having established the conceptual framework in this study, the development, and evaluation of the approach, the next step will be to try it out on an actual BIPV project proposal to aid/test practicality in real-life scenarios. This can be used to showcase the possible advantages and impact of the approach so it can be enhanced or reviewed accordingly. Also, a wider UX survey can be run with the actual approach to compare with the preliminary pilot, evaluate its effectiveness with potential clients, and evaluate stakeholder preferences. These two strategies are in view, and the authors hope to complete them in the near future.

Beyond the scope of this investigation, the discussion on BIPV advantages will continue within various disciplines as more studies validate its applicability towards net zero building [8,89]. Aspects of architectural integration, engineering, and manufacturing techniques as well as material technology will continue to develop. Research investigations show strategies for improved energy savings up to 10% via optimization in specific hot climate conditions [90], and improved power and performance ratio up to 40% via customization [91,92]. It is likely that investigations like these will continue to drive the rise of high performing BIPV systems in the coming years. However, communicating this technology and presenting its multi-dimensional benefits to all stakeholders will need to be advanced. The concept of educating the public towards advancing an appreciation of advantages will continue to be crucial in sustainability [93]. Within professional circles, university courses, professional development seminars, and communiqués on post-occupancy evaluations of BIPV projects may be used along with media campaigns for the general public to encourage adoption.

## 6. Conclusions

This study discussed the development of a conceptual framework for an educative-communication approach for BIPV market and research proposals. The paper examined various interacting topics in the BIPV literature relating to the importance of the BIPV hierarchy. It highlighted the main pillars of sustainability and applied this in the development of a communication approach. The research also discussed and presented its findings of a pilot UX survey and a checklist to evaluate the approach developed. From the existing body of literature, it is obvious there is no precise recipe for how to communicate the advantages of using solar energy in building design. In each project, the development team, project managers, and architects will have different dynamics, as well as varying or similar design objectives for each. Obsolete knowledge and a lack of understanding about aspects of solar energy, and information on older unappealing and nonfunctional designs need to be readdressed in the light of innovative products and technological development. Obstacles can be addressed through communication—accessible information, good examples, and ideal solutions—which emphasize research-based evidence, promote interest and knowledge sharing, in order to realize high-quality solar energy solutions in architecture [35].

From our findings, understanding environmental and economic benefits proves to be an important driver for people who know the benefits of using solar energy for electricity in homes and are likely adopters. We thus conclude that improving understanding of these benefits using the developed approach can potentially facilitate and encourage BIPV adoption. This study elaborates the need and strategies for appropriate dissemination of innovative ideas. The approach and findings can be applied in other contexts, and for other technological breakthroughs or new product development which advance a sustainable global future.

**Author Contributions:** "Initial Conceptualization" of the investigation was carried out by D.E.A., K.A.T.A., and A.H.; "Investigation and Methodology" was carried out by D.E.A. and T.O.A.; "Supervision" of the text was carried out by K.A.T.A. and A.H. "Statistical analysis" was carried out by T.O.A. The "Evaluation" section was developed and executed by S.O.A., D.E.A., and T.O.A.; "Writing" of the original draft was carried out by D.E.A. and T.O.A.; "Writing review and editing" was carried out by all the authors.

**Funding:** This research was funded in part by the UAE University (UAEU).

**Acknowledgments:** The authors appreciate the support given by the UAE University (UAEU).

**Conflicts of Interest:** The authors declare no conflict of interest.

## References

1. Jelle, B.P. Building integrated photovoltaics: A concise description of the current state of the art and possible research pathways. *Energies* **2016**, *9*, 21. [CrossRef]
2. Radhi, H. On the value of decentralised PV systems for the GCC residential sector. *Energy Policy* **2011**, *39*, 2020–2027. [CrossRef]
3. Hiremath, R.B.; Shikha, S.; Ravindranath, N.H. Decentralized energy planning; modeling and application—A review. *Renew. Sustain. Energy Rev.* **2007**, *11*, 729–752. [CrossRef]
4. Banos, R.; Manzano-Agugliaro, F.; Montoya, F.G.; Gil, C.; Alcayde, A.; Gómez, J. Optimization methods applied to renewable and sustainable energy: A review. *Renew. Sustain. Energy Rev.* **2011**, *15*, 1753–1766. [CrossRef]
5. Toledo, O.M.; Oliveira Filho, D.; Diniz, A.S.A.C. Distributed photovoltaic generation and energy storage systems: A review. *Renew. Sustain. Energy Rev.* **2010**, *14*, 506–511. [CrossRef]
6. Sauter, R.; Watson, J. Strategies for the deployment of micro-generation: Implications for social acceptance. *Energy Policy* **2007**, *35*, 2770–2779. [CrossRef]
7. Dunn, S.; Peterson, J.A. *Micropower: The Next Electrical Era*; Worldwatch Institute: Washington, DC, USA, 2000.
8. Kylili, A.; Fokaides, P.A. Investigation of building integrated photovoltaics potential in achieving the zero energy building target. *Indoor Built Environ.* **2014**, *23*, 92–106. [CrossRef]
9. Bonomo, P.; Chatzipanagi, A.; Frontini, F. Overview and analysis of current BIPV products: New criteria for supporting the technological transfer in the building sector. *VITRUVIO-Int. J. Arch. Technol. Sustain.* **2015**, 67–85. [CrossRef]
10. Designing Photovoltaic Systems for Architectural Integration. Available online: http://task41.iea-shc.org/data/sites/1/publications/task41A3-2-Designing-Photovoltaic-Systems-for-Architectural-Integration.pdf (accessed on 1 April 2018).
11. Montoro, D.F.; Vanbuggenhout, P.; Ciesielska, J. *Building Integrated Photovoltaics: An Overview of the Existing Products and Their Fields of Application*; Report Prepared in the Framework of the European Funded Project; SUNRISE: Saskatoon, SK, Canada, 2011.
12. Heinstein, P.; Ballif, C.; Perret-Aebi, L.-E. Building integrated photovoltaics (BIPV): Review, potentials, barriers and myths. *Green* **2013**, *3*, 125–156. [CrossRef]
13. Energy Systems in Architecture-Integration Criteria and Guidelines. Available online: https://infoscience.epfl.ch/record/197097/files/T41DA2-Solar-Energy-Systems-in-Architecture-28March20131.pdf (accessed on 1 April 2018).
14. Jelle, B.P.; Breivik, C.; Røkenes, H.D. Building integrated photovoltaic products: A state-of-the-art review and future research opportunities. *Sol. Energy Mater. Sol. Cells* **2012**, *100*, 69–96. [CrossRef]
15. Scognamiglio, A.; Farkas, K.; Frontini, F.; Maturi, L. Architectural quality and photovoltaic products. In Proceedings of the 27th European Photovoltaic Solar Energy Conference and Exhibition (EU PVSEC), Frankfurt, Germany, 24–28 September 2012; pp. 24–28.
16. Thomas, R. What are photovoltaics? In *Photovoltaics and Architecture*; Taylor & Francis: London, UK, 2003; pp. 18–28.

17. Morton, O. *Solar Energy: A New Day Dawning? Silicon Valley Sunrise*; Nature Publishing Group: London, UK, 2006.
18. Lewis, N.S.; Nocera, D.G. Powering the planet: Chemical challenges in solar energy utilization. *Proc. Natl. Acad. Sci. USA* **2006**, *103*, 15729–15735. [CrossRef] [PubMed]
19. Bakos, G.C.; Soursos, M.; Tsagas, N.F. Technoeconomic assessment of a building-integrated PV system for electrical energy saving in residential sector. *Energy Build.* **2003**, *35*, 757–762. [CrossRef]
20. Sharples, S.; Radhi, H. Assessing the technical and economic performance of building integrated photovoltaics and their value to the GCC society. *Renew. Energy* **2013**, *55*, 150–159. [CrossRef]
21. Timilsina, G.R.; Kurdgelashvili, L.; Narbel, P.A. Solar energy: Markets, economics and policies. *Renew. Sustain. Energy Rev.* **2012**, *16*, 449–465. [CrossRef]
22. Van Sark, W.G.J.H.M.; Arancon, S.; Weiss, I.; Tabakovic, M.; Fechner, H.; Louwen, A.; Georghiou, G.; Makrides, G. Development of BIPV courseware for students and professionals: The Dem4BIPV Project. In Proceedings of the 33rd European Photovoltaic Solar Energy Conference, Amsterdam, The Netherlands, 28 September 2017; pp. 2895–2899.
23. Sawin, J. *Renewable Energy Policy Network for the 21st Century Renewables 2017 Global Status Report*; REN21 Secretariat: Paris, France, 2017; pp. 1–302.
24. IEA PVPS. Snapshot of Global Photovoltaic Markets; Report IEA PVPS T1-31. 2017. Available online: http://www.iea-pvps.org/fileadmin/dam/public/report/statistics/IEA-PVPS_-_A_Snapshot_of_Global_PV_-_1992-2016__1_.pdf (accessed on 1 April 2018).
25. Ritzen, M.; Reijenga, T.; El Gammal, A.; Warneryd, M.; Sprenger, W.; Rose-Wilson, H.; Payet, J.; Morreau, V.; Boddaert, S. IEA-PVPS Task 15: Enabling Framework for BIPV Acceleration (IEA-PVPS). In Proceedings of the 48th IEA PVPS Executive Commitee Meeting, Vienna, Austria, 15–16 November 2016; Volume 16.
26. Prieto, A.; Knaack, U.; Auer, T.; Klein, T. Solar façades-Main barriers for widespread façade integration of solar technologies. *J. Façade Des. Eng.* **2017**, *5*, 51–62.
27. Tabakovic, M.; Fechner, H.; Van Sark, W.; Louwen, A.; Georghiou, G.; Makrides, G.; Loucaidou, E.; Ioannidou, M.; Weiss, I.; Arancon, S. Status and outlook for building integrated photovoltaics (BIPV) in relation to educational needs in the BIPV sector. *Energy Procedia* **2017**, *111*, 993–999. [CrossRef]
28. Goh, K.C.; Goh, H.H.; Yap, A.B.K.; Masrom, M.A.N.; Mohamed, S. Barriers and drivers of Malaysian BIPV application: Perspective of developers. *Procedia Eng.* **2017**, *180*, 1585–1595. [CrossRef]
29. Yang, R.J.; Zou, P.X. Building integrated photovoltaics (BIPV): Costs, benefits, risks, barriers and improvement strategy. *Int. J. Constr. Manag.* **2016**, *16*, 39–53. [CrossRef]
30. Karakaya, E.; Sriwannawit, P. Barriers to the adoption of photovoltaic systems: The state of the art. *Renew. Sustain. Energy Rev.* **2015**, *49*, 60–66. [CrossRef]
31. Yang, R.J. Overcoming technical barriers and risks in the application of building integrated photovoltaics (BIPV): Hardware and software strategies. *Autom. Constr.* **2015**, *51*, 92–102. [CrossRef]
32. Mousa, O. BIPV/BAPV Barriers to Adoption: Architects' Perspectives from Canada and the United States. Master's Thesis, University of Waterloo, Waterloo, ON, Canada, 2014.
33. Azadian, F.; Radzi, M.A.M. A general approach toward building integrated photovoltaic systems and its implementation barriers: A review. *Renew. Sustain. Energy Rev.* **2013**, *22*, 527–538. [CrossRef]
34. Koinegg, J.; Brudermann, T.; Posch, A.; Mrotzek, M. It Would Be a Shame if We Did Not Take Advantage of the Spirit of the Times.... An Analysis of Prospects and Barriers of Building Integrated Photovoltaics. *GAIA-Ecol. Perspect. Sci. Soc.* **2013**, *22*, 39–45. [CrossRef]
35. Probst, M.M.; Roecker, C. Criteria for architectural integration of active solar systems IEA Task 41, Subtask A. *Energy Procedia* **2012**, *30*, 1195–1204. [CrossRef]
36. Taleb, H.M.; Pitts, A.C. The potential to exploit use of building-integrated photovoltaics in countries of the Gulf Cooperation Council. *Renew. Energy* **2009**, *34*, 1092–1099. [CrossRef]
37. Attoye, D.E.; Tabet Aoul, K.A.; Hassan, A. A Review on Building Integrated Photovoltaic Façade Customization Potentials. *Sustainability* **2017**, *9*, 2287. [CrossRef]
38. International Renewable Energy Agency (IRENA). *Renewable Power Generation Costs in 2017*; IRENA: Abu Dhabi, United Arab Emirates, 2018.
39. Nejat, P.; Jomehzadeh, F.; Taheri, M.M.; Gohari, M.; Majid, M.Z.A. A global review of energy consumption, CO2 emissions and policy in the residential sector (with an overview of the top ten CO2 emitting countries). *Renew. Sustain. Energy Rev.* **2015**, *43*, 843–862. [CrossRef]

40. World Energy Council (WEC). *World Energy Resources 2013 Survey*; World Energy Council: London, UK, 2013.
41. A Call to Action: Buildings Key to Corporate Sustainability. Available online: https://www.environmentalleader.com/2008/09/a-call-to-action-buildings-key-to-corporate-sustainability/ (accessed on 20 August 2018).
42. Hagemann, I.B. Examples of successful architectural integration of PV: Germany. *Prog. Photovolt. Res. Appl.* **2004**, *12*, 461–470. [CrossRef]
43. Blessing, L.T.; Chakrabarti, A. *DRM, a Design Research Methodology*; Springer Science & Business Media: New York, NY, USA, 2009.
44. Yin, R.K. *Case Study Research and Applications: Design and Methods*; Sage Publications: London, UK, 2017.
45. Eisenhardt, K.M.; Graebner, M.E. Theory building from cases: Opportunities and challenges. *Acad. Manag. J.* **2007**, *50*, 25–32. [CrossRef]
46. Stigka, E.K.; Paravantis, J.A.; Mihalakakou, G.K. Social acceptance of renewable energy sources: A review of contingent valuation applications. *Renew. Sustain. Energy Rev.* **2014**, *32*, 100–106. [CrossRef]
47. Hansen, J.P.; Narbel, P.A.; Aksnes, D.L. Limits to growth in the renewable energy sector. *Renew. Sustain. Energy Rev.* **2017**, *70*, 769–774. [CrossRef]
48. Kabir, E.; Kumar, P.; Kumar, S.; Adelodun, A.A.; Kim, K.-H. Solar energy: Potential and future prospects. *Renew. Sustain. Energy Rev.* **2018**, *82*, 894–900. [CrossRef]
49. Hernandez, R.R.; Easter, S.B.; Murphy-Mariscal, M.L.; Maestre, F.T.; Tavassoli, M.; Allen, E.B.; Barrows, C.W.; Belnap, J.; Ochoa-Hueso, R.; Ravi, S. Environmental impacts of utility-scale solar energy. *Renew. Sustain. Energy Rev.* **2014**, *29*, 766–779. [CrossRef]
50. Singh, G.K. Solar power generation by PV (photovoltaic) technology: A review. *Energy* **2013**, *53*, 1–13. [CrossRef]
51. Kandpal, T.C.; Broman, L. Renewable energy education: A global status review. *Renew. Sustain. Energy Rev.* **2014**, *34*, 300–324. [CrossRef]
52. Strantzali, E.; Aravossis, K. Decision making in renewable energy investments: A review. *Renew. Sustain. Energy Rev.* **2016**, *55*, 885–898. [CrossRef]
53. Panwar, N.L.; Kaushik, S.C.; Kothari, S. Role of renewable energy sources in environmental protection: A review. *Renew. Sustain. Energy Rev.* **2011**, *15*, 1513–1524. [CrossRef]
54. Thirugnanasambandam, M.; Iniyan, S.; Goic, R. A review of solar thermal technologies. *Renew. Sustain. Energy Rev.* **2010**, *14*, 312–322. [CrossRef]
55. Bazilian, M.; Onyeji, I.; Liebreich, M.; MacGill, I.; Chase, J.; Shah, J.; Gielen, D.; Arent, D.; Landfear, D.; Zhengrong, S. Re-considering the economics of photovoltaic power. *Renew. Energy* **2013**, *53*, 329–338. [CrossRef]
56. Solangi, K.H.; Islam, M.R.; Saidur, R.; Rahim, N.A.; Fayaz, H. A review on global solar energy policy. *Renew. Sustain. Energy Rev.* **2011**, *15*, 2149–2163. [CrossRef]
57. Kannan, N.; Vakeesan, D. Solar energy for future world—A review. *Renew. Sustain. Energy Rev.* **2016**, *62*, 1092–1105. [CrossRef]
58. Khan, J.; Arsalan, M.H. Solar power technologies for sustainable electricity generation—A review. *Renew. Sustain. Energy Rev.* **2016**, *55*, 414–425. [CrossRef]
59. Hafez, A.Z.; Soliman, A.; El-Metwally, K.A.; Ismail, I.M. Tilt and azimuth angles in solar energy applications—A review. *Renew. Sustain. Energy Rev.* **2017**, *77*, 147–168. [CrossRef]
60. Lewis, N.S. Research opportunities to advance solar energy utilization. *Science* **2016**, *351*, aad1920. [CrossRef] [PubMed]
61. Sampaio, P.G.V.; González, M.O.A. Photovoltaic solar energy: Conceptual framework. *Renew. Sustain. Energy Rev.* **2017**, *74*, 590–601. [CrossRef]
62. Parida, B.; Iniyan, S.; Goic, R. A review of solar photovoltaic technologies. *Renew. Sustain. Energy Rev.* **2011**, *15*, 1625–1636. [CrossRef]
63. Jäger-Waldau, A. European Photovoltaics in world wide comparison. *J. Non-Cryst. Solids* **2006**, *352*, 1922–1927. [CrossRef]
64. Feltrin, A.; Freundlich, A. Material considerations for terawatt level deployment of photovoltaics. *Renew. Energy* **2008**, *33*, 180–185. [CrossRef]

65. Luthander, R.; Widén, J.; Nilsson, D.; Palm, J. Photovoltaic self-consumption in buildings: A review. *Appl. Energy* **2015**, *142*, 80–94. [CrossRef]
66. Zomer, CD.; Costa, M.R.; Nobre, A.; Rüther, R. Performance compromises of building-integrated and building-applied photovoltaics (BIPV and BAPV) in Brazilian airports. *Energy Build.* **2013**, *66*, 607–615. [CrossRef]
67. Sahin, I. Detailed review of Rogers' diffusion of innovations theory and educational technology-related studies based on Rogers' theory. *Turk. Online J. Educ. Technol.* **2006**, *5*, 14–23.
68. Rogers, E.M. *Diffusion of Innovations*, 5th ed.; A Division of Macmillan Publishing Co Inc.: New York, NY, USA; Free Press: New York, NY, USA, 2003.
69. Wisdom, J.P.; Chor, K.H.B.; Hoagwood, K.E.; Horwitz, S.M. Innovation adoption: A review of theories and constructs. *Adm. Policy Ment. Health Ment. Health Serv. Res.* **2014**, *41*, 480–502. [CrossRef] [PubMed]
70. Yin, R.K. Discovering the future of the case study. Method in evaluation research. *Eval. Pract.* **1994**, *15*, 283–290. [CrossRef]
71. Eisenhardt, K.M. Better stories and better constructs: The case for rigor and comparative logic. *Acad. Manag. Rev.* **1991**, *16*, 620–627. [CrossRef]
72. Shakir, M. *The Selection of Case Studies: Strategies and Their Applications to IS Implementation Case Studies*; Massey University: Palmerston North, New Zealand, 2002.
73. Siggelkow, N. Persuasion with case studies. *Acad. Manag. J.* **2007**, *50*, 20–24. [CrossRef]
74. The Communication Process. Available online: https://infoscience.epfl.ch/record/197100/files/T41C1-CommunicationsGuide-2012.pdf (accessed on 1 April 2018).
75. Espeche, J.M.; Noris, F.; Lennard, Z.; Challet, S.; Machado, M. PVSITES: Building-integrated photovoltaic technologies and systems for large-scale market deployment. *Multidiscip. Digit. Publ. Inst. Proc.* **2017**, *1*, 690. [CrossRef]
76. Femenías, P.; Thuvander, L.; Gustafsson, A.; Park, S.; Kovacs, P. Improving the market up-take of energy producing solar shading: A communication model to discuss preferences for architectural integration across different professions. In Proceedings of the 9th Nordic Conference on Construction Economics and Organization, Göteborg, Sweden, 13–14 June 2017; Volume 13, p. 140.
77. PVSITES Project. BIPV Demo Sites. Available online: http://www.pvsites.eu/project/demo-sites/ (accessed on 20 August 2018).
78. Parris, T.M.; Kates, R.W. Characterizing and measuring sustainable development. *Annu. Rev. Environ. Resour.* **2003**, *28*, 559–586. [CrossRef]
79. Beccali, M.; Cellura, M.; Mistretta, M. Decision-making in energy planning. Application of the Electre method at regional level for the diffusion of renewable energy technology. *Renew. Energy* **2003**, *28*, 2063–2087. [CrossRef]
80. Terrados, J.; Almonacid, G.; PeRez-Higueras, P. Proposal for a combined methodology for renewable energy planning. Application to a Spanish region. *Renew. Sustain. Energy Rev.* **2009**, *13*, 2022–2030. [CrossRef]
81. Disley, Y.P. Sustainable development goals for people and planet. *Nature* **2013**, *495*, 305–307.
82. Kumar, A.; Sah, B.; Singh, A.R.; Deng, Y.; He, X.; Kumar, P.; Bansal, R.C. A review of multi criteria decision making (MCDM) towards sustainable renewable energy development. *Renew. Sustain. Energy Rev.* **2017**, *69*, 596–609. [CrossRef]
83. The Future of Sustainability: Re-thinking Environment and Development in the Twenty-first Century. Available online: https://cmsdata.iucn.org/downloads/iucn_future_of_sustanability.pdf (accessed on 1 April 2018).
84. United Nations. World Summit Outcome. Available online: http://www.un.org/womenwatch/ods/A-RES-60-1-E.pdf (accessed on 1 April 2018).
85. Robert, K.W.; Parris, T.M.; Leiserowitz, A.A. What is sustainable development? Goals, indicators, values, and practice. *Environ. Sci. Policy Sustain. Dev.* **2005**, *47*, 8–21. [CrossRef]
86. Attoye, D.E.; Tabet Aoul, K.A.; Hassan, A. Development of A Building Integrated Photovoltaics—Mass Custom Housing. In Proceedings of the 5th Zero Energy Mass Custom Homes International Conference, Kuala Lumpur, Malaysia, 20–23 December 2016; pp. 142–152.
87. International Standards Organization. *ISO 9241-210: Ergonomics of Human System Interaction-Part 210: Human-Centred Design for Interactive Systems*; International Organization for Standardization (ISO): Geneva, Switzerland, 2008.

88. Hassenzahl, M.; Tractinsky, N. User experience-a research agenda. *Behav. Inf. Technol.* **2006**, *25*, 91–97. [CrossRef]
89. Baetens, R.; De Coninck, R.; Van Roy, J.; Verbruggen, B.; Driesen, J.; Helsen, L.; Saelens, D. Assessing electrical bottlenecks at feeder level for residential net zero-energy buildings by integrated system simulation. *Appl. Energy* **2012**, *96*, 74–83. [CrossRef]
90. Al Dakheel, J.; Tabet Aoul, K.; Hassan, A. Enhancing Green Building Rating of a School under the Hot Climate of UAE; Renewable Energy Application and System Integration. *Energies* **2018**, *11*, 2465. [CrossRef]
91. Nagy, Z.; Svetozarevic, B.; Jayathissa, P.; Begle, M.; Hofer, J.; Lydon, G.; Willmann, A.; Schlueter, A. The adaptive solar facade: From concept to prototypes. *Front. Arch. Res.* **2016**, *5*, 143–156. [CrossRef]
92. Valckenborg, R.M.E.; van der Wall, W.; Folkerts, W.; Hensen, J.L.M.; de Vries, A. Zigzag Structure in Façade Optimizes PV Yield While Aesthetics are Preserved. In Proceedings of the 32nd European Photovoltaic Solar Energy Conference and Exhibition, Munich, Germany, 20–24 June 2016; European Commission: Brussels, Belgium; pp. 647–650.
93. Adekunle, T.O. Autonomous Living: An Eco-social Perspective. *Int. J. Constr. Environ.* **2015**, *6*, 1–15. [CrossRef]

© 2018 by the authors. Licensee MDPI, Basel, Switzerland. This article is an open access article distributed under the terms and conditions of the Creative Commons Attribution (CC BY) license (http://creativecommons.org/licenses/by/4.0/).

Article

# Greenhouse Gas Emissions and Economic Performance in EU Agriculture: An Empirical Study in a Non-Linear Framework

Eleni Zafeiriou [1,*], Ioannis Mallidis [2], Konstantinos Galanopoulos [1] and Garyfallos Arabatzis [3]

1 Department of Agricultural Development, Democritus University of Thrace, GR68200 Orestiada, Greece; k_galan@otenet.gr
2 Department of Mechanical Engineering, Aristotle University of Thessaloniki, P.O. Box 461, 54124 Thessaloniki, Greece; imallidi@auth.gr
3 Department of Forestry Management of the Environment and Natural Resources, Democritus University of Thrace, GR68200 Orestiada, Greece; garamp@fmenr.duth.gr
* Correspondence: ezafeir@agro.duth.gr

Received: 16 August 2018; Accepted: 19 October 2018; Published: 23 October 2018

**Abstract:** Numerous linkages among Agriculture and climate change have been identified and validated in global terms. In European Union, the economic performance–carbon dioxide emission relationship has become a particularly high priority issue for Common agricultural policy within the last decade, attracting scientific interest. Within this socio–economic framework, the present work studies the relationship between agricultural carbon emissions equivalents and income per capita for the agricultural sector in different EU countries with the assistance of the nonlinear autoregressive distributed lag (NARDL) cointegration technique. Our findings validate the existence of a strong relationship between GHG emissions and agricultural income, since the cointegration among the two variables is established in all instances, while the asymmetric impact of agricultural income on carbon emissions may well provide policy makers with tools which when implemented, may well promote the increase of agricultural income along with GHG effect mitigation in a successful way.

**Keywords:** sustainable agriculture; negative externalities; GHG emissions; NARDL model

## 1. Introduction

The agricultural income–climate change relationship in global and European Union (EU) terms can be documented through different interlinkages while the concept of sustainability has become a high priority issue for EU agriculture [1]. The climate change–agriculture interaction is bi-directional [2]. Explicitly, climate change may affect global economic performance in the sector of agriculture through its impact on productivity [3], while on the other hand agriculture is a major contributor to global warming [1]. Other issues related to the climate change–agriculture interaction which has arisen within the last two decades include poverty and food security problems [3].

Furthermore, apart from the negative impacts, a number of positive effects of Greenhouse Gas emissions (GHG) to crop growth have also been recorded in recent literature. To be more specific, rising concentrations in the atmosphere improves agriculture in two different ways; first of all by stimulating photosynthesis and secondly by decreasing water requirements. However, the reaction of crops to carbon emissions respond depends on their physiology and other prevailing conditions such as water and nutrient availability, pests and diseases [4]. For the reasons described above, the positive impacts of the GHG effect for agriculture are expected to lead to a 16% increase in productivity corresponding to 45 million hectares in the cultivation of northern Europe focused especially on high latitudes [5] (Altieri et al., 2015).

Currently, GHG emissions from agriculture contribute with 10% to EU-28 total GHG emissions [6] (EC, 2018). This proportion is characterized by variability among EU member states. This variation is related to the size and importance of the agricultural sector in each individual country. Major concern regarding the issue of climate change has led to a decrease in GHG emissions by 23% within the last two decades. The pace through which the decrease occurs has been decreasing with a progressively limiting rate. The decreasing rate in EU GHG emissions in agriculture can be attributed to several factors, including among others an increase in productivity, decreases in cattle numbers, improvements in farm management practices and developments and implementation of agricultural and environmental policies [7] (Van Vuuren et al., 2017).

Among the solutions found for the limitation of the climate change impact of agriculture in terms of farming practices are the following [5,8] (Locatelli et al., 2015; Altieri and Nicholls 2017); (i) use of nitrification inhibitors to increase the efficiency of the nitrogen applied and at the same time reduce nitrous oxide emissions from mineral fertilizers; (ii) an efficient use of fertilization in terms of timing (iii) precision farming as a crop management concept to respond to inter- and intra-field variability in crops and last but certainly not least organic farming. A great size variation is observed not only in terms of the size of organic farming, but also in terms of growth in the total organic area. The typically best management practices that are used already may lead to lower farm emissions while also possibly leading to cost savings and increasing farm profitability. These measures provide major benefits, while the aforementioned options may be carried out at the farm level, or in some cases multiple farmers and other stakeholders ([9] Swinton et al., 2018).

Therefore, with the implementation of the particular practices we are raising agricultural productivity and income in the smallholder production sector, which in turn may drive economic transformation and growth in agriculture [10] (Schmidhuber and Tubiello 2007). Therefore, agriculture in terms of policy implementation may play a central role in the effort of climate change mitigation in the EU. Despite the fact that agriculture is a sector for which emission reduction is a priority, only a few EU countries have set quantitative targets for agriculture [11] (Fellman et al., 2018).

To synopsize, EU efforts focus on the reductions in GHG emissions without limiting the competitiveness of EU agriculture and its ability to satisfy growing global food demand, and these efforts are tightly related to agricultural income. Within this framework the present manuscript has as an objective to survey the relationship carbon emissions equivalent generated by agriculture–agricultural income in selected countries of European Union with variation in terms of the size of agriculture, of the point of economic growth as well as the behavior of the farmers regarding the implementation of specific environmental measures ([3,12] Frank et al., 2017; Kalfagianni and Kuik 2017).

In short, the main objectives of our research effort are the following;

- Detect the true nature of the relationship and the existence of non-linearities in the relationship among the variables.
- Capture asymmetric responses to positive and negative changes over both the short-and the long-run through positive and negative partial sum decompositions of the explanatory variables.
- Test the validity of the environmental Kuznets curve (ECK) hypothesis for the sector of agriculture for different EU countries.

The manuscript is organized as follows: The next section describes the existing literature, Section 3 presents the agriculture in different EU countries, Section 4 provides a subtle description of data used and the methodology on Nonlinear Autoregressive Distributed-lagged (NARDL) modelling, Section 4 outlines the results of the data process and an insight in the implications of those results, while the last section concludes.

## 2. Literature Review

The economic growth–environmental degradation relationship as synopsized in the environmental Kuznets curve (ECK) hypothesis has been a subject of extended study within the last couple of decades. Different methodologies and different indexes for environmental degradation have impeded the consensus on the validation of the environmental Kuznets curve hypothesis [13–18]. The interaction among environmental degradation and economic performance either in terms of a firm, a sector or economy has been mostly studied with the use of atmospheric indicators, while literature on the EKC hypothesis employs land indicators, like water ecosystems, biodiversity indicators, and freshwater indicators. The majority of the existing literature has argued that an inversed U-shape exists for the environmental degradation income per capita relationship, the steepness of which is mostly affected by income elasticity, scale, composition and technique effects, and international trade [19] (Sarkodie and Strezov 2019).

The majority of the empirical studies are based on the carbon emissions, since global anthropogenic greenhouse gas emissions are mainly attributed to carbon emissions [19]. Furthermore the carbon emissions are mostly generated by sectors related to energy use, forestry, agricultural processes and land use [20]. Regarding the methodologies used, a time series analysis with the assistance of cointegration linear or nonlinear conflict results was derived [2,13,21–24].

In the case panel data the most widespread methodologies are the Pedroni cointegration technique and Fully Modified Ordinary Least Squares (FMOLS) that validated the EKC hypothesis for developed countries [25], On the other hand, Ozcan (2013) [26] rejected the validity of the ECK hypothesis [18], examine the validity of the EKC hypothesis, with respect to the relationship between economic growth and environmental sustainability in Africa, while Javid & Sharif (2016) [27] propose an autoregressive distributed lag (ARDL) cointegration method in order to evaluate the impact of financial development, per capita real income, square of per capita real income, per capita energy consumption and openness on the per capita $CO_2$ emissions for Pakistan during 1972–2013. Finally, Culas (2007) [28] propose a panel data method, which incorporates both cross-sectional and time series data, in order to evaluate the impact of environmental policies on the EKC relationship for deforestation across Latin American, African and Asian countries.

Concerning the sector of agriculture, a few works can be mentioned with different econometric methodologies. For instance, Coderoni and Esposti [29] analyze the long-term relationship between agricultural GHG emissions and productivity growth in order to assess emissions sustainability for Italian regional agriculture. Another recent research involved the impact of renewable and agriculture on greenhouse gas emissions [30]. According to their findings, an increase in renewable energy and agriculture leads to a decrease in $CO_2$ emissions, while the opposite is validated for the case of non-renewable energy. Furthermore, in the short-run Granger causality is validated from non-renewable energy to emissions and to agriculture, from economic growth to agriculture, and from agriculture to renewable energy in a direct way, implying that sustainable agriculture may promote renewable energy and decrease carbon emissions [30] (Liu et al., 2017). Another work on agriculture based on annual data and with ARDL model by Zafeiriou and Azam (2017) and Zafeiriou et al. (2017) [2,21] concluded that the adoption of environment-friendly farming practices and crops' selection does not secure high economic and environmental performance simultaneously, at least in the short run, for our sample countries. Furthermore, in the long run the existing situation asks for the modification of the agro–environmental measures adopted to make those two targets complementary and not mutually exclusive for a farmer. To be more specific, the challenge of climate change mitigation in the sector of agriculture that remains is to effectively incorporate sustainability into the agricultural operations, management, research and development [31].

The existing literature as analyzed in the present section, indicate that even though the environmental Kuznets curve hypothesis is broadly examined in a linear and non-linear framework, no previous research effort can be found to our knowledge which implements the nonlinear autoregressive distributed lag (NARDL) cointegration method for evaluating the GHG emissions—economic performance relationship for the sector of agriculture. Thus, the present research effort employs for first time a nonlinear framework with the assistance of an NARDL model in order to confirm the existence causality and asymmetric effects of a relationship among carbon emissions and economic performance in EU agriculture.

## 3. The Agriculture in EU Countries

The purpose of this section is to describe the sector of agriculture for the countries examined in terms of the degree of mechanization of their agriculture, the percentage of people employed in the agricultural sector, the antiquity years of agricultural machinery use, and the greening of farming as reflected in the size of organic agriculture. This section may well provide us with plausible explanations for the results of the model analyzed in Section 4.

The selection of the countries for our study was based on: (i) the significance of the sector of agriculture for the economy (proportion to total GDP) (ii) their attitude towards the environmental friendly practices and (iii) whether they are old or new members of the EU.

### 3.1. Bulgaria

The first country examined is Bulgaria where only 8.04% of the employed in the agriculture sector are under the age of 35 years, and 33.59% are over 64 years old. Moreover, the educational level in Bulgarian agriculture is relatively low. More than half of the employees have primary and secondary education according to the National Institute of Statistics, and only 3% of farm managers have agricultural education. The CAP and its implementation in Bulgaria require specific qualifications: Knowledge in the fields of information technology, management, environmental practices, etc.

The majority of farms are marked by low mechanization. The low level of mechanization, its absence in some of the farms and the use of old equipment in most cases (over 85% of the used equipment is older than 10 years) involves the development of primitive, low-productive and inefficient production, which poses serious constraints to competitiveness. The status of agriculture in Bulgaria, as analyzed has shown that the future development and transformation of the agricultural sector into a competitive one has proven to be a highly complex and responsible task [32] (Kagatsume and Todorova 2007). This transformation is a requirement of both European and world markets. The structure of the agricultural sector that has been established in respect to the amount of utilized agricultural area (UAA) is abnormal and does not contribute to the development of the agrarian sector in Bulgaria [33] (Todorova 2016). In 2007, 54.1% of Bulgarian farms in size are smaller than 0.5 ha, and only 0.8 percent of the farms are over 100 ha in size, i.e., more than half the farms in the country cultivated only 1.5% of the total UAA. This reveals a structure of agriculture in which small farms predominate. The main reason for the existence of such a structure is the method of land restitution in its real boundaries which was adopted. The dimensional structure of farms in Bulgaria is the factor which most greatly restricts the creation of viable farms. The small farms that are prevailing are characterized by low profitability that cannot attract young people to become involved in agriculture [34] (Nikolova). The permanent establishment of semi-commercial farms is an inhibitory factor for the formation of market-oriented farms.

Regarding the size and growth of organic farming in Bulgaria the following can be mentioned; for the time period 2012–2016, Bulgaria is one of the countries studied, exhibiting a growth in the total organic area of over 100%. Furthermore, an interesting index for the potential growth in the organic sectors for the years to come is the area under conversion as a percentage of the total organic area, which for Bulgaria is equal to 77.5% on of the largest shares recorded for EU countries [35].

### 3.2. United Kingdom (UK)

In the UK, the Utilised Agricultural Area (UAA) has remained quite stable, as it only lost 63,250 hectares (−0.4%) between the two reference years: It covered 64% of the country's territory in 2010, the second highest share reported within the EU-28 after the one recorded in Ireland (71%).

The average area per holding was quite high (84 ha per farm) in 2010—it actually increased by about 5 hectares over the period under analysis, hence on average British farms were found to be the second largest within the EU-28, after Czech farms, which recorded a much larger average area (152 ha per farm). The number of people regularly working in the agriculture sector decreased (−18.8%) between 2000 and 2010, as about 97,000 people stopped working on the farms. Therefore, the labor force in the agricultural sector represented only 1.4% of the active British population in 2010, one of the lowest shares recorded among the EU Member States. Although the farm animal population decreased by 15.7% (−2.5 million LSU) for the period studied, the United Kingdom reported a value (about 13.3 million LSU) that exceeded the highest value respectively recorded within the EU-28 in 2010. Agriculture in UK is considered to be the most mechanized agricultural sector, given the large size and the increased prosperity of farming (Long. 1963). As for the labor force in this particular sector, 346,000 people were working on British farms in 2016, less than 1% of the total employment. Finally, the agricultural labor force as measured in annual work units (AWU) does not show a difference in the decrease, given that figures dropped by 22.3% from 317,280 AWU to 246,650 AWU. A last but certainly not least issue is the greening of UK agriculture as reflected to the adoption to organic farming. A downward trend is recorded for the United Kingdom (−16.9%), for the 2012–2016 period. Regarding the area under conversion as a percentage of the total organic area for the year 2016, the United Kingdom had a share of less than 10%, one of the smallest for the EU [36].

*3.3. Spain*

For the case of Spain in the year 2010 there were 989,800 agricultural holdings in Spain, a 23.1% drop compared to 2000 and much in line with the common trend recorded in most of the EU countries. In 2010 the UAA in Spain represented 47% of the whole territory; a decrease of 9.2% was reported when compared to the results of the previous census. In terms of the average size of the agricultural holdings, an increase of 18% was observed, shifting from an average of 20.3 ha in 2000 to 24.0 ha in 2010. The overall Spanish livestock, expressed in livestock units (LSU), only changed marginally and amounted to 14.8 million LSU in 2010, a 1% decrease when compared to 2000. In addition the employment in agriculture dropped by 8.7% between 2000 and 2010, passing from 2.4 million to 2.2 million. However, the population working in agriculture still represented 9.8% of the economically active population of Spain in 2010 [1]. According to the FSS 2010 data, there was an average of 0.52 hectares of UAA per inhabitant in Spain. This ratio indicates a decrease (−21%) compared to the one recorded during the Agricultural census in 2000, when the UAA per inhabitant was 0.66 hectares. This result is a combination of both a higher population (+14.8%) and a lower UAA (−9.2%). In addition, Spain represented one of the highest total organic areas not only in 2012 but also in the year 2016, reaching the figure of 16.9% of the total farming area [37].

*3.4. Greece*

The agricultural sector in Greece remains an important sector of economic activity and employment for Greece, with exports of agricultural products accounting for one third of total exports in Greece. Agriculture contributes to 4.1 percent of GDP and is characterized by small farms and low capital investment. Greece's utilized agricultural area is close to 5 million hectares, of which 57 percent is in the plains and 43 percent is in mountainous or semi-mountainous areas. There are about 150 million olive trees in the country, either in systematic orchards or scattered across the country. Lower agricultural productivity in Greece, compared to other EU Member States, is correlated to the smaller average-size of holdings. The economies of scale offered by modern farming practices have limited impact on the small plots of land typically used in Greece. Regarding the farming systems adopted in Greece and in particular organic farming presents a downward trend (-25.9%) to the already small size of the organic areas [38].

*3.5. France*

France is a dominant agricultural country in global terms. However officially, the share of population actively involved in farming is decreasing. On the other hand, new creative methods of marketing and agritourism have given a boost to the sector. In addition, almost half of farm income in France is generated by livestock raising, and the rest is contributed by crops. The GDP generated by agriculture (% of GDP) corresponds to 1.6407% in 2016, based on the World Bank database. In addition, the utilized agricultural area (UAA), remained quite stable within the last decade since it was only decreasing by 3.2. Therefore, agricultural land covered 43% of French national territory and was the largest across the whole EU-27 in 2010 (EU, 2018). Furthermore, the case of France represents one of the three highest total organic areas not only in 2012 but also in the year 2016, while the size of the organic area reaches 12.9% of the total farming area for the year 2016 [39].

*3.6. Germany*

Germany is one of the most important agricultural producers in the EU. To be more specific, Germany is second only to France in animal production and fourth following France, Italy and Spain, in vegetable production. In terms of employment, almost 10 percent of all of Germany's gainfully employed population work (2005) in the agricultural industry. However, this rate is decreasing due to modernization of the agricultural process. Furthermore, the agriculture, value added (% of GDP) in Germany was reported at 0.63432% in 2017 (World Bank data base). Germany is one of the countries with the highest size of organic area that along with Italy, France and Spain account for more than half organic land in the EU [40].

Having considered all the aforementioned issues, it would be of interest to examine the interrelationships among carbon emissions generated by agriculture and agricultural income in different EU countries with heterogeneity in the economic conditions of each country and also particularities in the farming systems. The major though common issue is the implementation of the Common Agricultural Policy and the specific agro–environmental measures as formatted within the last decade [18,19].

## 4. Data—Methodology

The present work uses annual data of Greece, France, Spain (all of them being Mediterranean countries) Bulgaria, a newly entrant country and two old member states with a well-organized farming sector: Germany and United Kingdom. The data employed were derived by FAOSTAT and the sample time period runs from 1970 to 2014. Implicitly, the bivariate framework being employed includes data on; carbon emissions equivalent ($CO_2t$) in thousands of tonnes, generated by agriculture per 1000 hectares of Utilized Agricultural Area (UAA), as proxy for environmental degradation and the net value added per capita ($NVAt$) generated by agriculture as proxy for agricultural income. The methodology employed is the nonlinear ARDL model (non linear cointegration) introduced by Reference [41] aiming to detect the existence of non linearities and asymmetric effects in the relationship among the variables studied.

The major advantages of the particular methodology are the estimation simplicity, the greater flexibility in relaxing the assumptions that the time-series should be integrated of the same order while it provides the potential to identify with accuracy the absence of cointegration, linear cointegration and nonlinear cointegration [42].

Prior to the implementation of the nonlinear cointegration (NARDL model) we implemented a break unit root test. Until now a number of different unit roots has been used such as the Augmented Dickey Fuller (ADF test), Phillips Perron (PP test), Elliot, Rothenberg, and Stock (ERS test), Ng and Perron (NP), and Kwiatkowski, Phillips, Schmidt, and Shin (KPSS) tests, with occasionally conflict results.

The conflict results may be attributed to the existence of structural change in a time series which in turn causes bias to the unit root test results. Due to this observation, a large quantity of literature has been developed outlining various unit root tests that remain valid in the presence of a break [43].

In the present manuscript we employ a modified augmented Dickey-Fuller test allowing for levels and trends that differ across a single break date [44,45]. Different types of the aforementioned methodologies were used and the results are provided in the results' section.

In most studies, when causality and linear cointegration confirm that the dependent variable is expected to respond in a symmetric way to increases and decreases of the independent variable we employ the linear unrestricted Error Correction Model as provided by Equation (1).

$$\Delta Cem_t = \mu + \rho_{Cem} CO_{2t-1} + \rho_x Y_{t-1} + \sum_{i=1}^{p-1} a_i \Delta CO_{2t-1} + \sum_{i=0}^{q-1} \beta_i \Delta Y_{t-1} + \varepsilon_t \quad (1)$$

In the present manuscript this is not the case and therefore we employ the NARDL model through which cointegration non linearities and causality are simultaneously detected. For that reason, a decomposition of the net value added (NVAt) into its positive and negative partial sums was preceded, while the generated sums are provided by the following equations;

$$NVA_t^+ = \sum_{j=1}^{t} \Delta NVA_j^+ = \sum_{j=1}^{t} \max(\Delta NVA_j, 0) \text{ and } Y_t^- = \sum_{j=1}^{t} \Delta NVA_j^- = \sum_{j=1}^{t} \min(\Delta NVA_j, 0)$$

Therefore, the Error correction Model is taking the following form when partial sums of the exogenous variable are taken into consideration.

$$\Delta Cem_t = \mu + \rho_{Cem} Cem_{t-1} + \theta^+ NVA^+{}_{t-1} + \theta^- NVA^-{}_{t-1} + \sum_{i=1}^{p-1} a_i \Delta Cem_{t-1} + \sum_{i=0}^{q-1} (\omega^+{}_i \Delta YNVA^+{}_{t-1} + \omega^-{}_i \Delta NVA^-_{t-1}) + \varepsilon_t \quad (2)$$

The superscripts (+) and (−) in Equation (2) stand for the positive and negative partial sums decomposition as defined above.

The symbols $p$ and $q$ denote the respective lag orders for the dependent variable and the exogenous variable in the distributed lag part, respectively. In particular, the long-run symmetry can be tested by using a Wald test of the null hypothesis that

$$\theta^+ = \theta^-$$

Computing the positive and negative long-run coefficients as follows:

$$L_{NVA^+} = -\frac{\theta^+}{\rho_{Cem}} \quad L_{NVA^-} = -\frac{\theta^-}{\rho_{Cem}}$$

The short-run adjustments to the positive and negative shocks affecting the level of the carbon emissions generated by agriculture, are captured by the aforementioned parameters respectively.

The long run (LR) and the short run (SR) symmetry can both be tested by using a Wald test, while the short-run adjustments to the positive and negative shocks affecting the dependent variable, are captured by the parameters $\omega^+{}_i$ and $\omega^-{}_i$ respectively.

Finally, the traditional (linear) ECM as mentioned above can be used if both null hypotheses of short-run and long-run symmetry cannot be rejected.

The rejection of either the long-run, or the short-run symmetry will lead to the estimation of a cointegrating NARDL model with SR & LR asymmetry as follows:

$$\Delta Cem_t = \mu + \rho_{Cem} Cem_{t-1} + \rho_y Y_{t-1} + \sum_{i=1}^{p-1} a_i \Delta Cem_{t-1} + \sum_{i=0}^{q-1} (\omega^+{}_i \Delta NVA^+{}_{t-1} + \omega^-{}_i \Delta NVA^-_{t-1}) + \varepsilon_t \quad (3)$$

$$\Delta Cem_{2t} = \mu + \rho_{CO2}Cem_{t-1} + \theta^{+}YNVA^{+}{}_{t-1} + \theta^{-}NVA_{t-1} + \sum_{i=1}^{p-1} a_i \Delta Cem_{t-1} + \sum_{i=0}^{q-1}(\omega_i \Delta NVA_{t-1}) + \varepsilon_t \quad (4)$$

In short, the NARDL model may estimate the short-run dynamics through the distributed lag part and the long-run dynamics through a single cointegrating vector. Asymmetries can be captured not only in the short run but also in the long run. And last but not least the cointegration test that applies to the unrestricted model is an F-test on the joint hypothesis that the coefficients of the lagged level variables are jointly equal to zero [46].

## 5. Results and Discussion

The first step in our effort to study the relationship among carbon emissions equivalent and per capita agricultural income is to exclude the existence of I(2) and higher degree of integration variables for each individual country in order to proceed to the second step. The results are provided in Table 1.

**Table 1.** Results of breakpoint ADF unit root test.

| Variables | Trend/Break Specification | T—Statistic | Critical Values (5%) | Break Date |
|---|---|---|---|---|
| | | Bulgaria | | |
| cembul | Both/trend | −2.446296 | −4.524826 | 2006 |
| NVA$_{bul}$ | Intercept/intercept | −1.444585 | −4.193627 | 1991 |
| | | France | | |
| cem$_{fr}$ | Both/Both | −3.653095 | 4.616123 | 1985 |
| NVA$_{FR}$ | Both/Both | −4.270830 | 4.616123 | 1999 |
| | | Greece | | |
| cem$_{gr}$ | Both/Both | −4.183147 | 4.616123 | 1985 |
| NVA$_{GR}$ | Both/Intercept | −2.906720 | −4.859812 | 1987 |
| | | Spain | | |
| cemsp | Both/Intercept | −4.603551 | −4.859812 | 2007 |
| NVA$_{SP}$ | Both/Intercept | −4.291036 | −4.859812 | 1988 |
| | | Germany | | |
| CEE$_{GER}$ | Both/Intercept | −4.439 | −4.8598 | 1991 |
| NVA$_{GER}$ | Intercept/Intercept | −3.509 | −4.443649 | 1985 |
| | | United Kingdom | | |
| CEE$_{uk}$ | Intercept/Intercept | −2.976 | −4.443649 | 1999 |
| NVA$_{UK}$ | Intercept/Intercept | −3.486451 | −4.443649 | 1986 |
| | | Bulgaria | | |
| Δcembul | Both/trend | −4.832198 ** | −4.524826 | 1993 |
| ΔNVA$_{bul}$ | Intercept/Intercept | −10.19594 *** | −4.734858 | 1990 |
| | | France | | |
| Δcemfr | Both/Both | −7.047084 *** | 4.616123 | 1987 |
| ΔNVA$_{FR}$ | Both/Both | −7.173298 *** | 4.616123 | 1995 |
| | | Greece | | |
| Δcemgr | Both/Both | −4.993328 ** | 4.616123 | 2009 |
| ΔNVA$_{GR}$ | Both/Intercept | −6.880485 ** | −4.859812 | 2003 |
| | | Spain | | |
| Δcemsp | Both/Intercept | −9.868212 *** | −4.859812 | 2005 |
| ΔNVA$_{SP}$ | Both/Intercept | −5.787173 *** | −4.859812 | 2003 |
| | | Germany | | |
| Δcem$_{GER}$ | Both/Intercept | −6.368473 *** | −4.859812 | 1992 |
| ΔNVA$_{GER}$ | Both/Intercept | −7.557033 *** | −4.443649 | 2011 |
| | | United Kingdom | | |
| Δcemuk | Intercept/Intercept | −6.186 *** | −4.443649 | 1996 |
| ΔNVA$_{UK}$ | Both/Intercept | −7.545339 *** | −4.443649 | 2009 |

** denotes reject of unit root hypothesis in 5% level of significance; *** denotes reject of unit root hypothesis in 1% level of significance.

According to the results derived all the time series surveyed are I(1) implicitly non stationary in levels but stationary at first differences. The particular result provides us with the potential to implement the NARDL methodology provided the non validation of the time series employed as I(2).

Another issue to be discussed concerning the results illustrated in Table 1 involves the breaks validated for the variables studied. To be more specific, for the case of carbon emissions equivalent and for the countries studied including Bulgaria, France, Greece, Spain, and United Kingdom the break was validated for the following dates; 2006, 1985, 1985, 2007, 1991, 1999 respectively. For the case of carbon emissions the existence of the structural breaks may well be attributed either to Kyoto Protocol, Paris agreement and efforts organized aiming at climate change mitigation. On the other hand and regarding the net value added for the sector of agriculture, the indicated break dates for the aforementioned countries are the following ones respectively; 1991, 1999, 1987, 1988, 1985, 1986. The most common factor for the existence of the validated structural breaks and in particular for the variable of the agricultural income involve Cap reform either in 1990 or Agenda 2000, while the small changes may be attributed either to news dissemination or as a subsequent result. Furthermore, as country specific conditions, for the case of UK the foot-and-mouth disease (1985) can be mentioned, as well the changes in the political situation in Bulgaria (1991).

In addition, it is interesting to make a thorough analysis to the break date confirmed by the break unit root tests as provided in Table 1.

The next step in our analysis includes testing the existence on long or short run asymmetries in the behavior of the relationship among carbon emissions equivalent generated by agriculture and agricultural income.

The results of confirmation regarding the long run and short run asymmetry with the assistance of the Wald test for all the sample countries are provided in Table 2.

**Table 2.** Results of the long run and short run symmetry tests for the sample EU countries (carbon emissions as dependent variable).

| Pair of Variables | Long Run $W_{LR}$ | Short Run $W_{SR}$ | Conclusion |
|---|---|---|---|
| Cembul–NVA$_{BUL/cap}$ | 2.019015 * (0.0522) | 1.729280 * (0.0952) | NARDL with LR and SR asymmetry |
| Cemfr–NVA$_{FRL/cap}$ | 8.634360 ** (0.0004) | 9.645271 ** (0.0001) | NARDL with LR and SR asymmetry |
| Cemgr–NVA$_{GR/cap}$ | −1.638810 (0.1186) | −2.806545 ** (0.0121) | NARDL with LR symmetry and SR asymmetry |
| Cemsp–NVA$_{SP/cap}$ | 1.987999 * (0.0596) | −1.996244 * (0.0574) | NARDL with LR and SR asymmetry |
| CEE$_{GER}$–NVA$_{GER/cap}$ | −3.083814 ** (0.0071) | 3.084209 ** (0.0071) | NARDL with LR and SR asymmetry |
| CEE$_{UK}$–NVAUK$_{P/cap}$ | 1.9527 * (0.098) | 7.166741 ** (0.0004) | NARDL with LR and SR asymmetry |

* denotes reject of null hypothesis in 10% level of significance, ** denotes reject of null hypothesis in 1% level of significance. The parentheses denote the $p$-values.

The existence of dummy variables in the formation and evolution of the particular relationship has seemingly played a pivotal role, since a number of events can be recorded such as the entrance of each individual country to the EU, the fall of ex socialist political institutions as well as changes and reforms in the Common Agricultural Policy, including among others the greater concern regarding issues of GHG emissions mitigation (the introduction of agro–environmental policy measures with a major focus on the adoption of greening farming practice and satisfaction of sustainability criteria, as well as the extensive use of alternative energy sources in the farming process (including bioenergy and others).

For all the reasons mentioned above, detecting the existence and validating dummy variables is a very important step in the methodology process.

In order to detect the existence of the dummy variables, we initially estimate the NARDL model without using any dummy variables and with the existence of Bai Perron tests, we trace potential structural breaks. To be more specific, we test the null hypothesis of no breaks against a specific number of breaks (Bai and Perron 2003). Actually, the particular tests and their simulations' findings provide useful tools to the researchers through model selection processes and for the construction of confidence intervals for the break dates (in case multiple breaks exist) [45].

Initially we have to mention that the NARDL model for the case of Bulgaria validated the role of structural breaks as pivotal, with the assistance of Bai Perron process. Explicitly, the significance of impulse dummy variables confirms the impact of events on the formation of agricultural income: The first dummy variable validated as statistically significant corresponds to a structural break in 1986. Within that year poor harvests in 1985 and 1986 have been recorded, which in turn have led to grain imports of 1.8 and 1.5 million tons, respectively. The second dummy variable 1991 carried through the privatization of agricultural land may be attributed either to (a) the Assembly of European Fruit and Vegetable Growing and Horticultural Regions (AREFLH) supported the initiatives in favor of European regulation or to (b) the development of the renewable energy expansion in 1991. In a similar way to the other post-socialist regimes in eastern Europe, Bulgaria found the transition to capitalism more difficult than expected. In the year 2002 the privatization of agricultural land occurred, while in the same year AREFLH encouraged the initiatives supporting European regulation and in April 2013 the Assembly introduced the first Guidelines of European Practices in Integrated Production.

The major objective of developing sustainable farming is to produce and consume safe and quality food, and also to use available resources in an environmentally responsible manner, reducing costs, and minimizing the impact on the environment. It was also to promote the development of renewable energy expansion, particularly in the years 2008 to 2010. Within this framework in the next Table 3 we estimated the NARDL model for the agriculture of Bulgaria.

Table 3. NARDL estimation results for Bulgaria.

| Dependent Variable D ($Cem_{BUL}$) | | |
|---|---|---|
| Variables | Coefficient | Standard Error |
| C | 0.721248 *** | 0.111285 |
| $Cem_{BUL}$ (−1) | −0.332101 *** | 0.050969 |
| $NVA^+_{BUL}$ (−1) | 0.465118 *** | 0.147957 |
| $NVA^-_{BUL}$ (−1) | 0.854085 *** | 0.165330 |
| S_2002 | 0.133215 ** | 0.060634 |
| $D\ NVA^-_{BUL}$ | 0.594969 *** | 0.129314 |
| $D\ Cem_{BUL}$ (−1) | 0.544918 *** | 0.131120 |
| $D\ NVA^+_{BUL}$ (−3) | 0.411743 ** | 0.173579 |
| D_2002 | 0.133215 ** | 0.060634 |
| $D\ NVA^-_{BUL}$ (−4) | 0.251278 * | 0.122623 |
| $W_{LR}$ | 2.019015 * (0.0522) | |
| $L^+\ w$ | 1.40053 *** (0.0000) | |
| $L^-\ w$ | 2.5776 *** (0.000) | |
| $W_{SR}$ | 1.729280 ** (0.0952) | |
| Pss | 49.32123 *** (0.00000) | |
| ARCH | 0.947372 (0.3034) | |
| BG | 0.801545 (0.6698) | |

* denotes reject of null hypothesis in 10% level of significance, ** denotes reject of null hypothesis in 5% level of significance, *** denotes reject of null hypothesis in 1% level of significance, D-year denotes dummy variable, while S-year denotes seasonal dummy variable. The parentheses denote the p-values.

According to the results derived cointegration among the variables is confirmed, while no problems of heteroscedasticity or autocorrelation is validated. Finally, the positive sign of the partial decomposition independent variables is illustrative of a positive impact on carbon emissions generated by agriculture in case a change (either positive or negative change) in agricultural income occurs. Asymmetry in the long term as well as in the short term is validated while the sign of the relationship confirms the non-validity in Kuznets environmental curve.

Regarding France as the second country, being ranked as the European Union's largest producer and the second largest exporter has been a strong motivation for the selection of that particular country in the sample of the present work. However, this competitiveness has been put into peril within the last decade due to a number of factors, including the following (Faostat and World Bank Statistics):

The dismantling of the Common Agricultural Policy that had a strong impact on French and European farmers due to a stronger exposure to price volatility; The emergence of large agricultural powerhouses, such as Brazil, China or India, which have the ability to maintain their prices by social or environmental dumping; Last but certainly not least, is the loss of competitiveness of French agriculture within the European Union, due to fiscal and social harmonization, as well as higher employment costs in France.

Within this general framework we employed the nonlinear ARDL methodology, the results of which are provided in Table 4.

Table 4. NARDL estimation results for France.

| Dependent Variable D ($Cem_{FR}$) | | |
|---|---|---|
| | Coefficient | Standard Error |
| C | 1.364 | 0.341 |
| $CEM_{FR}$ (−1) | −0.479 | 0.123 |
| $NVA_{FR}^+$ (−1) | −0.265 | 0.0964 |
| $NVA_{FR}^-$ (−1) | −0.208 | 0.106 |
| @TREND | −0.0042 | 0.001 |
| D $NVA_{FR}^+$ (−5) | 0.158 | 0.087 |
| D $NVA_{FR}^-$ (−7) | −0.233 | 0.101 |
| D $NVA_{FR}^+$ (−6) | 0.189 | 0.0934 |
| D $NVA_{FR}^+$ (−4) | 0.163 | 0.090 |
| D $NVA_{FR}^-$ (−5) | −0.216 | 0.120 |
| $W_{LR}$ | −2.31 ** (0.0344) | |
| $L^+$ w | −0.553 *** (0.000) | |
| $L^-$ w | −0.434 *** (0.000) | |
| $W_{SR}$ | 3.212 *** (0.00370) | |
| Pss | 53.07 *** (0.00000) | |
| ARCH | 0.74 (0.39) | |
| Qlb(12) | 8.1809 (0.611) | |

* denotes reject of null hypothesis in 10% level of significance, ** denotes reject of null hypothesis in 5% level of significance, *** denotes reject of null hypothesis in 1% level of significance. The parentheses denote the *p*-values.

According to the results derived, we may see that only the time trend is found to be statistically significant in terms of time variables, cointegration among the variables surveyed is validated, while no Arch effects have been detected. An important issue that has to be underlined is that the negative impact of a change in agricultural income leads to a decrease in carbon emissions. Evidently, this result is indicative of validity for the Environmental Kuznets curve hypothesis, a result that is in line with that of Iwata et al. (2010) that also confirmed the validity of EKC for the whole economy taking into consideration also the role of nuclear energy in carbon emissions mitigation.

Another Mediterranean country that is also used in our sample that has a strong agriculture sector in terms of the EU is Spain. The estimation results of the NARDL model for the sector of agriculture in Spain are provided in Table 5.

Cointegration is validated for the data employed and for the case of agriculture according to our findings a result that is in line with the findings of previous studies on Spain though in terms of a whole economy. To be more specific previous studies confirmed that per capita GDP and $CO_2$ emissions are non-linearly cointegrated, providing support for the existence of the EKC hypothesis in Spain [47,48]. Furthermore, no heteroscedasticity or autocorrelation is validated while the impact of changes in agricultural income leads to a negative sign change in carbon emissions. Also asymmetry is validated in the long run as well as in the short run.

**Table 5.** NARDL estimation results for Spain.

| Dependent Variable D (CEESP) | | |
|---|---|---|
| | Coefficient | Standard Error |
| C | −0.256 *** | 0.0665 |
| CEESP (−1) | 0.392 *** | 0.0776 |
| NVA +SP (−1) | −0.267 *** | 0.0415 |
| NVA −SP (−1) | −0.213 *** | 0.058 |
| D CEESP (−1) | −0.827 *** | 0.125 |
| S_2003 | −0.057 ** | 0.027 |
| D NVA +SP (−1) | 0.186 ** | 0.086796 |
| S_1998 | 0.124 *** | 0.031 |
| S_1997 | 0.064 ** | 0.0267 |
| D CEESP (−2) | −0.406 *** | 0.107 |
| D NVA −SP (−1) | 0.402 *** | 0.132 |
| D_2007 | −0.0532 * | 0.0287 |
| D NVA −SP (−5) | 0.240 * | 0.127 |
| WLR | 1.988 * (0.0596) | |
| $L^+$ w | −0.680 *** (0.0000) | |
| $L^-$ w | −0.5430 *** (0.0000) | |
| WSR | −1.997 * (0.0574) | |
| Pss | 48.67 *** (0.000) | |
| ARCH | 0.0059 (0.9388) | |
| BG | 1.893 (0.3882) | |

\* denotes reject of null hypothesis in 10% level of significance, ** denotes reject of null hypothesis in 5% level of significance, *** denotes reject of null hypothesis in 1% level of significance D-year denotes dummy variable, while S-year denotes seasonal dummy variable. The parentheses denote the *p*-values.

Greece is another sample country for which agriculture does play a significant role in the economy, with changes occurring within the last decades since the tertiary sector including tourism, is becoming the main contributor to the Greek economy. The results derived for the case of Greece are provided in Table 6.

**Table 6.** NARDL estimation results for Greece.

| Dependent Variable D ($CEE_{GR}$) | Coefficient | Standard Error |
|---|---|---|
| C | 2.599 *** | 0.255 |
| $CEE_{GR}$ (−1) | −2.396 *** | 0.239 |
| $NVA_{GR}$ (−1) | −0.0835 *** | 0.019 |
| @TREND | −0.0032 *** | 0.0009 |
| S_1978 | −0.073 *** | 0.0124 |
| S_2009 | −0.063 *** | 0.018 |
| D $NVA_{GR}$- (−6) | 0.431 *** | 0.05 |
| D $NVA\ ^-_{GR}$ (−1) | 0.3271 *** | 0.052 |
| D $NVA_{GR}^+$ (−3) | 0.177 *** | 0.0278 |
| D $NVA_{GR}^+$ (−6) | −0.1292 ** | 0.0587 |
| D $CEE_{GR}$ (−4) | 1.029 *** | 0.1558 |
| D $CEE_{GR}$ (−1) | 1.496 *** | 0.192 |
| D $CEE_{GR}$ (−3) | 0.945 *** | 0.186 |
| D $NVA^-\ _{GR}$ | 0.0804 *** | 0.0273 |
| D $NVA_{GR}^+$ (−4) | −0.126 ** | 0.051 |
| D $NVA^-$ (−4) | 0.183 *** | 0.0376 |
| D $NVA^-$ (−2) | 0.232 *** | 0.055 |
| D $CEE_{GR}$ (−2) | 0.967 *** | 0.22 |
| $W_{LR}$ | −1.639 (0.119) | |
| L w | 0.0349 *** (0.0000) | |
| $W_{SR}$ | −2.8065 *** (0.0121) | |
| Pss | 167.35 *** (0.000) | |
| ARCH | 0.0594 (0.8163) | |
| Qlb(12) | 15.823 (0.226) | |

\* denotes reject of null hypothesis in 10% level of significance, ** denotes reject of null hypothesis in 5% level of significance, *** denotes reject of null hypothesis in 1% level of significance D-year denotes dummy variable, while S-year denotes seasonal dummy variable. The parentheses denote the *p*-values.

According to our findings, for the model estimated, no problems of heteroscedasticity or autocorrelation have been detected. Furthermore, asymmetry in the short run but not in the long run is validated for the case of Greece. The initiation of economic crisis in 2009 and the entrance of Greece in the EU seems to have played a statistically significant role in the formation of the carbon emissions–agricultural income relationship. Furthermore, a negative relationship among agricultural income and environmental degradation in terms of carbon emissions equivalent generated by agriculture is validated, and therefore the aforementioned process does not provide us with clear results on the validation of the EKC hypothesis.

Last but certainly not least, the sample includes recorded findings for two strong economies: Germany as an old member in the European Union and the UK which has recently become a non-member of the EU. Both have strong agricultural sectors that are adopting environmental friendly practices.

The results for the aforementioned countries Germany and United Kingdom are provided in the next Tables 7 and 8 respectively.

Table 7. NARDL estimation results for Germany.

| | Dependent Variable D ($CEE_{GER}$) | |
|---|---|---|
| | Coefficient | Standard Error |
| C | 1.601 *** | 0.332 |
| $NVA^-_{GER}(-1)$ | −1.031 *** | 0.188 |
| $NVA^+_{GER}(-1)$ | −0.78 *** | 0.155 |
| $CEE_{GER}(-1)$ | −0.226 *** | 0.060 |
| D_1992 | −0.368 *** | 0.059 |
| @TREND | −0.023 *** | 0.006 |
| $D\,CEE_{GER}(-7)$ | −0.136 | 0.104 |
| $D\,NVA^-_{GER}(-1)$ | 0.298 * | 0.151 |
| $D\,CEE_{GER}(-1)$ | 0.113 | 0.083 |
| $D\,NVA^+_{GER}(-5)$ | 0.464 *** | 0.181 |
| $D\,CEE_{GER}(-6)$ | −0.223 | 0.140 |
| $D\,NVA^+_{GER}(-1)$ | 0.586 ** | 0.187 |
| $D\,NVA^-_{GER}(-5)$ | −1.007 *** | 0.201 |
| $D\,CEE_{GER}(-5)$ | −0.983 *** | 0.174 |
| $D\,NVA^-_{GER}$ | 0.112 ** | 0.046 |
| $D\,NVA^+_{GER}(-4)$ | 1.168 *** | 0.197 |
| $D\,NVA^-_{GER}(-2)$ | 0.497 ** | 0.186 |
| $D\,CEE_{GER}(-3)$ | −0.822 *** | 0.141 |
| $D\,NVA^-_{GER}(-6)$ | −1.155 *** | 0.274 |
| D_1986 | −0.155 ** | 0.064 |
| $D\,NVA^-_{GER}(-4)$ | −0.4395 ** | 0.176 |
| D_1998 | −0.202 ** | 0.069 |
| $D\,NVA^+_{GER}(-6)$ | 0.3497 * | 0.163 |
| $D\,NVA^+_{GER}(-3)$ | 0.517 ** | 0.159 |
| D_2004 | 0.315 *** | 0.059 |
| $D\,NVA^+_{GER}$ | −0.956 *** | 0.203 |
| $W_{LR}$ | 3.419159 * (0.0644) | |
| $L^+\,w$ | −3.451 *** (0.000) | |
| $L^-\,w$ | −4.561 *** (0.000) | |
| $W_{SR}$ | 19.404 *** (0.000) | |
| Pss | 12.24735 *** (0.0016) | |
| ARCH | 5.751 (0.1244) | |
| BG | 6.710857 * (0.0763) | |

* denotes reject of null hypothesis in 10% level of significance, ** denotes reject of null hypothesis in 5% level of significance, *** denotes reject of null hypothesis in 1% level of significance D-year denotes dummy variable, while S-year denotes seasonal dummy variable The parentheses denote the *p*-values.

**Table 8.** NARDL Estimation results for the United Kingdom.

| Variable | Coefficient | Std. Error |
| --- | --- | --- |
| C | 1.386358 ** | 0.379667 |
| $NVA^-_{UK}(-1)$ | −1.460268 *** | 0.311756 |
| $NVA^+_{UK}(-1)$ | −0.347307 ** | 0.137026 |
| $CEE_{UK}(-1)$ | −0.335171 * | 0.138687 |
| $DCEE_{UK}(-4)$ | 0.522842 ** | 0.154422 |
| D_2006 | −0.382427 *** | 0.077645 |
| $DCEE_{UK}(-3)$ | 0.906940 ** | 0.155460 |
| $D\,NVA^-_{UK}(-5)$ | 1.804910 *** | 0.389180 |
| $D\,NVA^+_{UK}(-8)$ | 0.766567 *** | 0.151332 |
| $D\,NVA^+_{UK}(-6)$ | 0.389050 ** | 0.128983 |
| $D\,NVA^+_{UK}(-7)$ | −1.550402 *** | 0.334756 |
| $DCEE_{UK}(-5)$ | 1.279123 *** | 0.229187 |
| $D\,NVA^+_{UK}(-1)$ | −1.116430 *** | 0.198264 |
| $D\,NVA^+_{UK}$ | 0.399833 ** | 0.136188 |
| D_2000 | 0.636952 *** | 0.103028 |
| $NVA^-_{UK}(-2)$ | 1.936357 *** | 0.366177 |
| $DCEE_{UK}(-8)$ | −0.786072 *** | 0.179954 |
| $NVA^-_{UK}(-4)$ | 4.126965 *** | 0.546447 |
| $DCEE_{UK}(-1)$ | 0.275924 | 0.170779 |
| $D\,NVA^+_{UK}(-3)$ | −0.302843 | 0.238409 |
| $D\,NVA^-_{UK}(-1)$ | 2.853299 *** | 0.440836 |
| $D\,NVA^-_{UK}$ | 0.333163 | 0.200081 |
| $D\,NVA^+_{UK}(-4)$ | −1.309466 *** | 0.271996 |
| D_1993 | −0.179180 ** | 0.066434 |
| $D\,NVA^+_{UK}(-2)$ | 1.61081 *** | 0.320704 |
| $DCEE_{UK}(-6)$ | −1.366 *** | 0.266191 |
| $D\,NVA^-_{UK}(-7)$ | 5.135320 *** | 0.776614 |
| $D\,NVA^-_{UK}(-3)$ | 2.338461 *** | 0.343106 |
| $W_{LR}$ | | |
| $L^+\,w$ | 1.036737 ***(0.000) | |
| $L^-\,w$ | 4.358 ***(0.000) | |
| $W_{SR}$ | 51.362 ***(0.000) | |
| Pss | 69.88 ***(0.000) | |
| ARCH | 0.694(0.8746) | |
| BG | 3.659 (0.16) | |

* denotes reject of null hypothesis in 10% level of significance, ** denotes reject of null hypothesis in 5% level of significance, *** denotes reject of null hypothesis in 1% level of significance D-year denotes dummy variable, while S-year denotes seasonal dummy variable. The parentheses denote the *p*-values.

According to the results derived for the case of Germany, nonlinear cointegration among the variables employed is established and asymmetric adjustment in the long run as well as in the short run is validated. Finally, no problems of serial correlation or heteroscedasticity have been detected. Also, for the case of Germany, a number of structural breaks was validated including 1986, 1992, 1998, 2004 as statistically significant.

The last country, United Kingdom was selected as a sample country because it has a highly mechanized agricultural sector with a stable utilized agricultural area (UAA). Furthermore, the average size of a UK holding is 81 ha, which is significantly higher than much of the rest of Europe, including countries such as France and Germany, according to statistics gathered by the EU.

The estimation of NARDL model has provided us with the following results presented in Table 8.

According to the results presented above nonlinear cointegration is validated, with no problems of serial correlation or heteroscedasticity, asymmetry in the long run as well as in the short run. Furthermore, the agricultural income evidently affects the carbon emissions in a negative way, that is an increase in agricultural income leads to a decrease in carbon emissions.

To synopsize, similar results were derived for all the sample countries with the exception of the countries of Greece, and Bulgaria. Therefore the NARDL methodology has provided conflict results regarding the validity of environmental Kuznets curve for all the countries with a well-developed agricultural sector and medium to high agricultural income.

## 6. Conclusions

The present manuscript surveys the behavior of carbon emissions equivalent generated by agriculture as a function of agricultural income per capita for the same sector. According to our findings, cointegration among the two variables is established in all instances and asymmetric impact of agricultural income on carbon emissions equivalent (millions of tonnes) per 1000 hectares is validated in most cases. This particular result provides evidence of a strong relationship among GHG emissions and agricultural income. Therefore the non-linear relationship and the fact that the impact of positive and negative changes is not of the same magnitude are validated for our data. With the exception of the case of Bulgaria, variations in agricultural income result in a decrease of carbon emissions generated by agriculture. Furthermore, if carbon emissions increase the agricultural income increases at a far greater margin than if carbon emissions decrease. This particular result is more than evident in the case of Spain. This result provides us with an indication of a non sustainable agriculture. Furthermore, another important finding concerns Greece, where carbon emissions decrease more significantly during negative shocks compared to the positive ones.

The statistical significance of impulse dummy variables is indicative of the impact of CAP reforms and the adoption of energy policy tools on the formation of agricultural income as well as for GHG emissions mitigation. Implicitly, in most cases, crucial milestones (CAP reforms) appear to have had a negative impact on income. However, the effect of EU policies on GHG emissions mitigation is somewhat mixed, a fact that stresses the need for more effective policy tools in order to secure sustainable economic growth. In addition, the present survey provided some evidence of the true relationship among agricultural income and carbon emissions, characterized by the existence of nonlinearities. Finally, with the result showing that the two variables are highly cointegrated, effective policy tools may promote the increase of agricultural income along with GHG effect mitigation, if the complex nature is accounted for a number of issues including; differences among Member States, different responses in the short run and the long run, asymmetric responses in positive and negative shocks and many others. However, the measures taken to limit GHG emissions from agricultural sector may vary in cost-effectiveness and practicality. To be more specific, measures that should be taken in order to reduce $CO_2$ emissions from soils or to enhance carbon sequestration involve the maintenance of permanent pasture, conservation tillage, appropriate crop rotation and cover crops [35].

The measures designed in order to be successful must be in line with the Circular Economy (CE) Package recently adopted by European Union. Therefore, for an environmentally friendly and sustainable agriculture industry, the measures taken should aim towards the limitation of the resource use, waste reduction and promotion of sustainable production and consumption [49]. The concept of a circular economy plays a key role for the sector of agriculture in order to enable the improvement of environmental quality, economic prosperity and social equity to be accomplished, for current and future generations. This strategy will assist the agricultural firms to mitigate the GHG effect through the limitation of negative externalities, securing eco–efficiency [36,50].

**Author Contributions:** Author Contributions: E.Z. and I.M. conceived and designed the econometric analysis. E.Z., G.A. and K.G. wrote the paper. All the authors analyzed the data.

**Funding:** This research received no external funding.

**Conflicts of Interest:** The authors declare no conflicts of interest.

## References

1. Lipper, L.; Thornton, P.; Campbell, B.M.; Baedeker, T.; Braimoh, A.; Bwalya, M.; Caron, P.; Cattaneo, A.; Garrity, D.; Henry, K.; et al. Climate-smart agriculture for food security. *Nat. Clim. Chang.* **2014**, *4*, 1068–1072. [CrossRef]
2. Zafeiriou, E.; Sofios, S.; Partalidou, X. Environmental Kuznets curve for EU agriculture: Empirical evidence from new entrant EU countries. *Environ. Sci. Pollut. Res.* **2017**, *24*, 15510–15520. [CrossRef] [PubMed]
3. Frank, S.; Havlík, P.; Soussana, J.F.; Levesque, A.; Valin, H.; Wollenberg, E.; Kleinwechter, U.; Fricko, O.; Gusti, M.; Herrero, M.; et al. Reducing GHG emissions in agriculture without compromising food security? *Environ. Res. Lett.* **2017**, *12*, 105004. [CrossRef]
4. Long, S.P.; Ainsworth, E.A.; Leakey, A.D.; Nösberger, J.; Ort, D.R. Food for thought: Lower-than-expected crop yield stimulation with rising $CO_2$ concentrations. *Science* **2006**, *312*, 1918–1921. [CrossRef] [PubMed]
5. Altieri, M.A.; Nicholls, C.I.; Henao, A.; Lana, M.A. Agroecology and the design of climate change-resilient farming systems. *Agron. Sustain. Dev.* **2015**, *35*, 869–890. [CrossRef]
6. European Comission. Trends in GHG Emissions. 2018. Available online: http://ec.europa.eu/eurostat/statistics-explained/index.php/Greenhouse_gas_emission_statistics#Trends_in_greenhouse_gas_emissions (accessed on 10 August 2018).
7. Van Vuuren, D.P.; Stehfest, E.; Gernaat, D.E.; Doelman, J.C.; Van den Berg, M.; Harmsen, M.; de Boer, H.S.; Bouwman, L.F.; Daioglou, V.; Edelenbosch, O.Y.; et al. Energy, land-use and GHG emissions trajectories under a green growth paradigm. *Glob. Environ. Chang.* **2017**, *42*, 237–250. [CrossRef]
8. Locatelli, B.; Pavageau, C.; Pramova, E.; Di Gregorio, M. Integrating climate change mitigation and adaptation in agriculture and forestry: Opportunities and trade-offs. *Wiley Interdiscip. Rev. Clim. Chang.* **2015**, *6*, 585–598. [CrossRef]
9. Swinton, S.M.; Rector, N.; Robertson, G.P.; Jolejole-Foreman, C.B.; Lupi, F. Farmer decisions about adopting environmentally beneficial practices. *Ecol. Agric. Landsc.* **2015**, 340–359.
10. Schmidhuber, J.; Tubiello, F.N. Global food security under climate change. *Proc. Natl. Acad. Sci. USA* **2007**, *104*, 19703–19708. [CrossRef] [PubMed]
11. Fellmann, T.; Witzke, P.; Weiss, F.; Van Doorslaer, B.; Drabik, D.; Huck, I.; Salputra, G.; Jansson, T.; Leip, A. Major challenges of integrating agriculture into climate change mitigation policy frameworks. *Mitig. Adapt. Strateg. Glob. Chang.* **2018**, *23*, 451–468. [CrossRef] [PubMed]
12. Kalfagianni, A.; Kuik, O. Seeking optimality in climate change agri-food policies: Stakeholder perspectives from Western Europe. *Clim. Policy* **2017**, *17*, 72–92. [CrossRef]
13. Olale, E.; Ochuodho, T.O.; Lantz, V.; El Armali, J. The environmental Kuznets curve model for GHG emissions in Canada. *J. Clean. Prod.* **2018**, *184*, 859–868. [CrossRef]
14. Xu, T. Investigating Environmental Kuznets Curve in China—Aggregation bias and policy implications. *Energy Policy* **2018**, *114*, 315–322. [CrossRef]
15. Su, E.C.-Y.; Chen, Y.-T. Policy or income to affect the generation of medical wastes: An application of environmental Kuznets curve by using Taiwan as an example. *J. Clean. Prod.* **2018**, *188*, 489–496. [CrossRef]
16. Balaguer, J.; Cantavella, M. The role of education in the Environmental Kuznets Curve. Evidence from Australian data. *Energy Econ.* **2018**, *70*, 289–296. [CrossRef]
17. Zhang, J.; Luo, M.; Cao, S. How deep is China's environmental Kuznets curve? An analysis based on ecological restoration under the Grain for Green program. *Land Use Policy* **2018**, *70*, 647–653. [CrossRef]
18. Lin, B.; Omoju, O.E.; Nwakeze, N.M.; Okonkwo, J.U.; Megbowon, E.T. Is the environmental Kuznets curve hypothesis a sound basis for environmental policy in Africa? *J. Clean. Prod.* **2016**, *133*, 712–724. [CrossRef]
19. Sarkodie, S.A.; Strezov, V. A review on Environmental Kuznets Curve hypothesis using bibliometric and meta-analysis. *Sci. Total Environ.* **2018**, *649*, 128–145. [CrossRef] [PubMed]
20. DiSano, J. *Indicators of Sustainable Development: Guidelines and Methodologies*; United Nations Department of Economic and Social Affairs, United Nations: New York, NY, USA, 2002.
21. Zafeiriou, E.; Azam, M. $CO_2$ emissions and economic performance in EU agriculture: Some evidence from Mediterranean countries. *Ecol. Indic.* **2017**, *81*, 104–114. [CrossRef]
22. Steinkraus, A. *Investigating the Carbon Leakage Effect on the Environmental Kuznets Curve Using Luminosity Data*; Economics Department Working Paper Series No. 15; Institut für Volkswirtschaftslehre: Braunschweig, Germany, 2016.

23. Gupta, V.; Yadav, U. Combining Indicators of Energy Consumption and $CO_2$ Emissions: EKC in India. *Int. J. Ecol. Econ. Stat.* **2016**, *37*, 56–74.
24. Katz, D. Water use and economic growth: Reconsidering the Environmental Kuznets Curve relationship. *J. Clean. Prod.* **2015**, *88*, 205–213. [CrossRef]
25. Apergis, N. Environmental Kuznets curves: New evidence on both panel and country-level $CO_2$ emissions. *Energy Econ.* **2016**, *54*, 263–271. [CrossRef]
26. Ozcan, B. The nexus between carbon emissions, energy consumption and economic growth in Middle East countries: A panel data analysis. *Energy Policy* **2013**, *62*, 1138–1147. [CrossRef]
27. Javid, M.; Sharif, F. Environmental Kuznets curve and financial development in Pakistan. *Renew. Sustain. Energy Rev.* **2016**, *54*, 406–414. [CrossRef]
28. Culas, R.J. Deforestation and the environmental Kuznets curve: An institutional perspective. *Ecol. Econ.* **2007**, *61*, 429–437. [CrossRef]
29. Coderoni, S.; Esposti, R. Is there a long-term relationship between agricultural GHG emissions and productivity growth? A dynamic panel data approach. *Environ. Resour. Econ.* **2014**, *58*, 273–302. [CrossRef]
30. Liu, X.; Zhang, S.; Bae, J. The impact of renewable energy and agriculture on carbon dioxide emissions: Investigating the environmental Kuznets curve in four selected ASEAN countries. *J. Clean. Prod.* **2017**, *164*, 1239–1247. [CrossRef]
31. Lozano, F.J.; Freire, P.; Guillén-Gozalbez, G.; Jiménez-Gonzalez, C.; Sakao, T.; Mac Dowell, N.; Ortiz, M.G.; Trianni, A.; Carpenter, A.; Viveros-García, T. New perspectives for sustainable resource and energy use, management and transformation: Approaches from green and sustainable chemistry and engineering. *J. Clean. Prod.* **2016**, *30*, 1–3. [CrossRef]
32. Kagatsume, M.; Todorova, S. *Impact of the EU Common Agricultural Policy on Farming Structure in Bulgaria*; Special Issue; Agricultural Economics Society of Japan: Tokyo, Japan, 2007; pp. 575–582.
33. Todorova, S. Bulgarian agriculture in the conditions of the EU Common Agricultural Policy. *J. Cent. Eur. Agric.* **2016**, *17*, 107–118. [CrossRef]
34. Nikolova, M. Relationship between the sustainable models of production in agriculture and the challenges to their development in Bulgaria. *J. Econ.* **2015**, *3*, 57–68. [CrossRef]
35. EC. 2018. Available online: https://ec.europa.eu/eurostat/statistics-explained/index.php/Agricultural_census_in_Bulgaria#Database (accessed on 30 July 2018).
36. EC. 2018. Available online: https://ec.europa.eu/eurostat/statistics-explained/index.php/Agricultural_census_in_UK#Database (accessed on 30 July 2018).
37. EC. 2018. Available online: https://ec.europa.eu/eurostat/statistics-explained/index.php/Agricultural_census_in_Spain#Database (accessed on 30 July 2018).
38. EC. 2018. Available online: https://ec.europa.eu/eurostat/statistics-explained/index.php/Agricultural_census_in_Greece#Database (accessed on 30 July2018 ).
39. EC. 2018. Available online: https://ec.europa.eu/eurostat/statistics-explained/index.php/Agricultural_census_in_France#Database (accessed on 20 July 2018).
40. EC. 2018. Available online: https://ec.europa.eu/eurostat/statistics-explained/index.php/Agricultural_census_in_Germany#Database (accessed on 30 June 2018).
41. Shin, Y.; Yu, B.; Greenwood-Nimmo, M. Modelling asymmetric cointegration and dynamic multipliers in a nonlinear ARDL framework. In *Festschrift in Honor of Peter Schmidt*; Springer: New York, NY, USA, 2014; pp. 281–314.
42. Constantinos, K.; Eleni, Z.; Nikolaos, S.; Bantis, D. GHG emissions–crude oil prices: An empirical investigation in a nonlinear framework. In *Environment, Development and Sustainability*; Springer: Dordrecht, The Netherlands, 2018; pp. 1–22.
43. Hansen, B.E. The new econometrics of structural change: Dating breaks in US labour productivity. *J. Econ. Perspect.* **2001**, *15*, 117–128. [CrossRef]
44. Narayan, P.K.; Popp, S. A new unit root test with two structural breaks in level and slope at unknown time. *J. Appl. Stat.* **2010**, *37*, 1425–1438. [CrossRef]
45. Bai, J.; Perron, P. Computation and analysis of multiple structural change models. *J. Appl. Econ.* **2003**, *18*, 1–22. [CrossRef]
46. Badeeb, R.A.; Lean, H.H. Asymmetric impact of oil price on Islamic sectoral stocks. *Energy Econ.* **2018**, *71*, 128–139. [CrossRef]

47. Sephton, P.; Mann, J. Further evidence of an environmental Kuznets curve in Spain. *Energy Econ.* **2013**, *36*, 177–181. [CrossRef]
48. Esteve, V.; Tamarit, C. Threshold cointegration and nonlinear adjustment between $CO_2$ and income: The environmental Kuznets curve in Spain, 1857–2007. *Energy Econ.* **2012**, *34*, 2148–2156. [CrossRef]
49. EC. 2017. Available online: https://ec.europa.eu/clima/policies/strategies/2020_en (accessed on 20 June 2018).
50. Núñez-Cacho, P.; Molina-Moreno, V.; Corpas-Iglesias, F.A.; Cortés-García, F.J. Family Businesses in Transition to a Circular Economy Model: The Case of "Mercadona". *Sustainability* **2018**, *10*, 538.

© 2018 by the authors. Licensee MDPI, Basel, Switzerland. This article is an open access article distributed under the terms and conditions of the Creative Commons Attribution (CC BY) license (http://creativecommons.org/licenses/by/4.0/).

Article

# Benchmarking Internet Promotion of Renewable Energy Enterprises: Is Sustainability Present?

## Zacharoula Andreopoulou * and Christiana Koliouska

Laboratory of Forest Informatics, Faculty of Forestry and Natural Environment, Aristotle University of Thessaloniki, P.O. Box 247, 54124 Thessaloniki, Greece; ckolious@for.auth.gr
* Correspondence: randreop@for.auth.gr; Tel.: +30-694-718-5206

Received: 24 September 2018; Accepted: 9 November 2018; Published: 14 November 2018

**Abstract:** Sustainability constitutes a broad discipline that focuses on the social, economic and environmental impact of human activities. Many policies and strategies have been developed for the pursuit of environmental sustainability and the guidance to a green society. Many enterprises have taken meaningful steps to improve their own environmental performance through corporate sustainability and environmental management. Environmental management contributes to significant improvements to environmental performance of the enterprises. This paper aims to evaluate the Renewable Energy Enterprises performance in the Internet in Thessaloniki Prefecture regarding the characteristics of sustainability using Multi-criteria Decision Analysis. TOPSIS method was used to provide a ranking of the Renewable Energy Enterprises according to their sustainability and finally conclude to a benchmark. According to the results of the research, the Renewable Energy Enterprises achieve a good level of sustainability but not the optimum. However, the entrepreneurs should adopt modern environmental policy, sustainable marketing, green network framework and certified environmental management system in order to consider their enterprise sustainable.

**Keywords:** Renewable Energy Enterprises; sustainability; Internet; benchmark

## 1. Introduction

Sustainability-related issues address many significant topics such as environment, energy, ecology, management, marketing, economics, research and development, transportation [1]. Sustainability science is situated as a science in which the societal values form the scientific agenda and at the same time, it provides both theoretical and practical knowledge to the society [2]. The vital role of environmental sustainability excellences is recognized as the organizations that belong to the third sector (such as the business world, public administration and civil society) have already adopted technical solutions against this background and which are behavioural examples of guidance to the green society [3,4]. Sustainability science has received far and away the most attention worldwide, due to the growing environmental problems and socioeconomic inequity, concluding to the current Global Economy Model (GEM), which emphasized profits [5]. Studies regarding the transdisciplinary collaboration indicate that there is progress in linking and incorporating the knowledge with action to support the sustainable use of natural resources, the climate change adaptation, the research agenda, decision making and the governance [6–12].

The proliferation of environmental sustainability-related policies during the last decades introduced a great interest in their functioning as tools of governance and their role in influencing environmental outcomes [13]. Sustainability issues are characterized as wicked problems that require cooperation among different parties in order to be defined and addressed [14]. Corporate sustainability has gone mainstream as many enterprises have already taken meaningful and important steps to enhance their own environmental performance. But while Corporate Political Actions (APC) such as lobbying can make a greater impact on environmental quality, they are frequently disregarded in most sustainability metrics and indices [15]. Whether forced by the concern for society and the environment, government regulation, stakeholder pressures, or economic profit, managers and strategists should continue to make important changes to achieve more efficient management of their socio-economic and environmental impacts—and to stay up to speed with the emerging market [16].

The ecologically sensitive corporate orientation sometimes referred to as the 'green' strategy, can originate with an enterprise's estimation of present-day production and marketing practices and adapting behaviour to indicate to a high level of environmental awareness [17]. A sustainable champion is defined as "the enterprise that has taken the lead in reducing the environmental impact of its activities, usually at levels beyond regulatory compliance and has achieved recognition as being 'green' compared with its competitors" [18,19]. The modern economic growth introduces new methods of organization and management, not only on national level but also on the levels of different economic entities, as well as on the replacement of the cumbersome technologies with the eco-friendly ones [20,21]. Sustainability innovations—new services or goods serving environmental and socio-economic goals [22–24]—constitute a critical attribute for many sectors (such as solar cell technology, electric cars, biofuels, biotechnology, bio-based plastics, wind-farms) but also provoke great uncertainty and ambiguity to the entrepreneurs and the intrapreneurs [25].

Many programs and initiatives have been established to help Small and Medium Enterprises (SMEs) to enhance their environmental performance, such as the Environmental Compliance Assistance Program for SMEs (ECAP) and the Green Action Plan (GAP), because SMEs seem to face more difficulties to conceive and implement environmental regulation [26,27]. Renewable energy support policies include research grants, development and demonstration projects, tax incentives for investment, fiscal and financial incentives and price-based and quantity-based policies such as feed-in tariffs, feed-in premiums, net metering, Renewable Energy Certificates (RECs), Renewable Portfolio Standards (RPSs) and competitive procurement for goods and services [28]. European Energy Industry is in the process of great revolution, which brings green power closer and can define its profiles for years to come [29]. Increasing the energy efficiency of large enterprises and SMEs plays a vital role in mitigating climate change [30], which is reflected in the EU energy efficiency target of 30% 2030 (Directive 2012/27/EU).

Energy enterprises ought to assume responsibility for improving the environment, beginning with the most fundamental practices such as minimizing, restoring and repairing the damaged environment in a timely fashion [31]. According to Bloomberg [32], investments in the sector of renewable energy declined by 8% in developed economies, while increased by 19% in developing economies [33].

There is a need to build a greener future, where technology, Internet of Things (IoT) and the economy will be replaced with green technology, green IoT and the green economy, respectively, which follows from a whole world of possible remarkable improvements of human welfare and therefore, supports the development of a smarter world [34]. ICT (Information and Communication Technology) integration and eco-innovation contribute not only to the main body of knowledge on sustainable marketing but also to the application of sustainable marketing amongst enterprises in developing economies [35]. Green ICT gains significant interest and it is considered as an important issue for the forthcoming years, while the enterprises are trying to compete with each other in how much "green" they are [36,37]. These modern green technologies provide significant opportunities for the people to advance in all areas [38]. Enterprises have been adapting their goods and services be more environmentally friendly [39,40]. Internet enables the collaboration with a variety of different enterprises, helping them to get sustainable competitive advantages in the global economic environment [41]. In particular, Internet is an effective channel for promoting an enterprise's green initiatives directly to consumers [42].

Nowadays, scientists, policymakers, managers and entrepreneurs are trying to find out how to turn IoT into reality and touch every aspect of our lives, since many technological constraints (such as standardization, interoperability, privacy and security issues, heterogeneity and data deluge) complicate the development of an IoT network and the transition to a smarter future [43]. The performance of a renewable energy enterprise through the Internet is totally affected by the sustainability awareness of the enterprise. According to a recent study [44], organizational agility affects positively the green performance of the enterprise, which positively affects customer satisfaction and organizational innovation. Furthermore, it is pointed out that the strong market orientation of an enterprise is an essential factor for high environmental performance of the enterprise [45]. The application of collaborative governance contributes to the achievement of sustainable benefits for the enterprise (e.g., creating technology legitimacy for sustainability, preventing food waste and enhancing environmental performance and compliance) [46]. The incorporation of sustainability into business practices through the implementation of sustainability programs lead to higher economic profit through eco-friendly innovative products [47].

For the purpose of this benchmarking study, we assess the Internet performance of the Renewable Energy SMEs in the Internet located in Thessaloniki Prefecture regarding their characteristics of sustainability Multi-criteria Decision Analysis. TOPSIS method was applied for the ranking of the Renewable Energy SMEs according to their sustainability and finally conclude to a benchmark.

## 2. Materials and Methods

The Internet presences of the Renewable Energy SMEs in Thessaloniki Prefecture are retrieved from the Internet through large-scale hyper textual search engines (such as "Google" "Yahoo" and "Bing") and thematic search engines from June to August 2018.

As for the characteristics of sustainability that were examined, they are suggested by Kernel [48] and Andreopoulou et al. [4] to evaluate the sustainability of an enterprise. However, only 8 of these characteristics of sustainability were selected to study in order to describe the current situation in Thessaloniki Prefecture (Table 1). Each characteristic is represented by a variable $X_i$. The first step was to implement quantitative analysis through a 2-dimentional table in order to examine the presence or absence of these criteria. The value of 0 and the value 1 were attributed to the variables $X_1$, $X_2$, $X_3$, $X_4$, $X_5$, $X_6$, $X_7$ and $X_8$ for the non-existence and the existence of each characteristic respectively.

Table 1. Characteristics of sustainability.

| Characteristic | Variable |
| --- | --- |
| Make environmental policy | $X_1$ |
| Eco-friendly tips | $X_2$ |
| Develop green shopping policy | $X_3$ |
| Information on green services and activities | $X_4$ |
| Involvement in local green networks | $X_5$ |
| Green success stories | $X_6$ |
| Implement certified environmental management system compatible with ISO or EMAS | $X_7$ |
| Make a review of important environmental impacts | $X_8$ |

Variable $X_1$ refers to the environmental policy of the enterprise as it is an important channel to market the environmental advances of the enterprises, while environmental advances constitute a concrete manifestation for enterprises including the integration of environmental regulation and social responsibility principles [49]. Variable $X_2$ is associated with the provision of eco-friendly guidelines and tips for sustainable living to online visitors (such as investment in eco-friendly technology, 3R policy—reduce reuse and recycle, building insulation), while variable $X_3$ represents the development of green shopping policy (e.g., provide online shopping, sell electrical appliances with Grade 1 Energy Efficiency Label and saving energy, sell products with minimal packaging, reuse the packaging materials). Variable $X_4$ refers to the provision of information on green services and activities and variable $X_5$ refers to the involvement of the enterprise in local green entrepreneurial networks aiming to increase the effectiveness of their business activities regarding the environmental protection. Variable $X_6$ represents the list of successful examples of sustainable applications in order to give inspiration and ideas. Variable $X_7$ is associated with the adoption of the most robust environmental management tool EMAS and the compliance with the ISO requirements, while variable $X_8$ deals with the existence of reviews using indicators and important environmental impacts such as impacts on climate change, acidification, ozone depletion, air pollution, chemical pollution, freshwater use, forest resources and so forth.

Since the characteristics of sustainability are partially or completely incompatible and by nature very distinct and measures in different units, the evaluations of subjective probabilities, the multi-criteria decision analysis (MCDA) is the method that fits better in evaluating sustainability of management model [50]. TOPSIS method, which was developed by Hwang and Yoon [51], is a broadly used multi-criteria method for improving the decision-making process. By using TOPSIS method, the decision-maker solves selection/evaluation problem because it is based on a sound logic, which represents the rational of human choice [52].

The main idea of TOPSIS method comes from the concept that the selected alternative should be closer to the Positive Ideal Solution (PIS) and further from the Negative Ideal Solution (NIS) [53–55]. PIS is called the solution that maximizes the benefit criteria and minimizes the cost criteria, whereas NIS is the solution that minimizes the benefit criteria and maximizes the cost criteria [56]. Although, two "reference" points are introduced in that method, the relative importance of the distances from these two points is not taken into consideration [57]. To sum up, the alternative optimal solution is the alternative with the minimum distance from the PIS and the maximum distance from the NIS [56]. The weights of criteria weights in the TOPSIS method are defined a priori [58]. Even though TOPSIS uses crisp numerical values to present the performance rating of alternatives and the criteria weights, the preferences of the decision makers are often abstract and cannot be represented in this way in reality [59]. Alternatives are ranked according to the value of their Closeness Coefficient (CC) in decreasing order, which is calculated regarding the distance of the respective alternative from both PIS and NIS [55]. CC takes a value between 0 and 1.

The procedure of TOPSIS method includes the following steps [60]:

- construction of normalized decision matrix
- construction of weighted normalized decision matrix
- selection of the PIS and NIS
- computation of separation measures and CC
- ranking of the alternatives.

TOPSIS method uses all the attribute information, presents the total ranking of the alternatives, while the given attribute preferences may be either dependent or independent [61–64]. TOPSIS method was applied in this case because it is the best-developed method in this field of multicriteria decision-making problems with simple computation process and high flexibility [65]. Furthermore, there are the following four main reasons [66,67]: (a) TOPSIS logic is rational and understandable; (b) the computation processes are straightforward; (c) this approach presents the best alternatives for each criterion through a mathematical formula; (d) the weights of the criteria are integrated into the procedures for comparison. In this case study, the weight of the criteria is the same (0.125).

## 3. Results

### 3.1. Statistics

The research on the Internet about the Renewable Energy SMEs in Thessaloniki Prefecture resulted in the retrieve of 23 Internet presences. In particular the internet research results are presented in Table A1. The achievement of each one of the characteristics of sustainability is presented in Figure 1. Almost all the Renewable Energy SMEs (91%) fulfil the fourth characteristic of sustainability regarding the provision of information on their green services and activities ($X_4$). Many SMEs provide a thematic about eco-friendly tips ($X_2$) and their green success stories ($X_6$) (48% and 61% respectively) while the

65% of them develops green shopping policy ($X_3$). Only the 26% is involved in local green networks ($X_5$) and the 30% implements certified environmental management system ($X_7$).

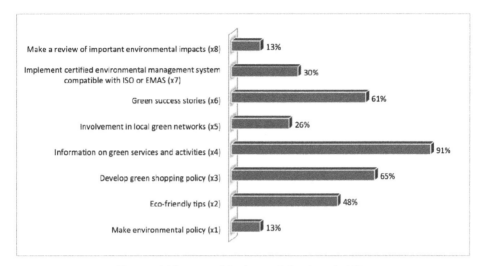

**Figure 1.** Achievement of the characteristics of sustainability.

Based on the application of the TOPSIS method, the total ranking of the Renewable Energy SMEs in Thessaloniki Prefecture according to their characteristics of sustainability retrieved from their Internet presence is presented in Table 2. The CC is estimated for each enterprise and it is used for the total ranking, as each enterprise with a higher CC is considered superior in ranking. According to these findings, the values estimated for CC present a spectrum of values between 0.19736 and 0.66281 and that indicates a great difference between the first and the last case in the ranking of the enterprises. The Renewable Energy SME with the best CC (REEnt_6) shows compliance with the legislation about environmental policy ($X_1$) (e.g., process control through standard operating procedures, environmental impact assessment of new projects, development and maintenance of constructive relationships with administration and local authorities, development of emergency response plan, environmental assessment policy and assessment of the environmental awareness of the suppliers), develops green shopping policy ($X_3$) through selling energy-efficient appliances, provides information on green services ($X_4$), describes some green success stories—case studies regarding the renewable energy development ($X_6$), implements certified environmental management system ISO 14001:2015 ($X_7$) and provides Environmental Impact Assessment Review ($X_8$). This enterprise can be used as a benchmark for the rest enterprises with lower CC. On the other side, the website of the Renewable Energy Enterprise with the worst CC (REEnt_14) provides only an overview of green success stories regarding the construction of solar parks.

The enterprises are further classified in two groups according to their ranking in order to present their level of sustainability. The average CC of the case is 0.3658. So, 12 enterprises belong to the group of "high sustainability" with average CC 0.4634 and the rest of the cases (11) belong to the group of "low sustainability" with average CC 0.2594. The averages of these two groups that were selected independently of each other was examined by using $t$-test for Independent Samples (with two options) in order to verify whether those group averages differ enough to believe that the enterprises from which they were selected have different averages. According to the results, there is clear differentiation between these two groups, as we reject the null hypothesis and accept the alternative one (Table 3).

**Table 2.** Total ranking of the Renewable Energy small medium enterprises (SMEs) according to their characteristics of sustainability.

|   | Renewable Energy Enterprises | Number of Achieved Characteristics | $d_i+$ | $d_i-$ | $CC_i$ |
|---|---|---|---|---|---|
| 1 | REEnt_6 | 6/8 | 0.06344 | 0.1247 | 0.66281 |
| 2 | REEnt_19 | 6/8 | 0.07906 | 0.11543 | 0.59352 |
| 3 | REEnt_9 | 6/8 | 0.08626 | 0.11016 | 0.56084 |
| 4 | REEnt_21 | 4/8 | 0.10529 | 0.09214 | 0.46668 |
| 5 | REEnt_13 | 5/8 | 0.10704 | 0.09009 | 0.45701 |
| 6 | REEnt_23 | 4/8 | 0.10708 | 0.09006 | 0.45683 |
| 7 | REEnt_7 | 5/8 | 0.10739 | 0.08968 | 0.45507 |
| 8 | REEnt_2 | 4/8 | 0.11732 | 0.07623 | 0.39383 |
| 8 | REEnt_11 | 4/8 | 0.11732 | 0.07623 | 0.39383 |
| 9 | REEnt_1 | 4/8 | 0.12017 | 0.07165 | 0.37353 |
| 9 | REEnt_3 | 4/8 | 0.12017 | 0.07165 | 0.37353 |
| 9 | REEnt_20 | 4/8 | 0.12017 | 0.07165 | 0.37353 |
| 10 | REEnt_17 | 3/8 | 0.12765 | 0.05728 | 0.30972 |
| 10 | REEnt_18 | 3/8 | 0.12765 | 0.05728 | 0.30972 |
| 11 | REEnt_15 | 3/8 | 0.12913 | 0.05387 | 0.29437 |
| 12 | REEnt_12 | 2/8 | 0.13082 | 0.04962 | 0.275 |
| 13 | REEnt_16 | 2/8 | 0.13195 | 0.04652 | 0.26068 |
| 14 | REEnt_8 | 2/8 | 0.1331 | 0.04313 | 0.24474 |
| 15 | REEnt_4 | 2/8 | 0.13338 | 0.04226 | 0.2406 |
| 15 | REEnt_5 | 2/8 | 0.13338 | 0.04226 | 0.2406 |
| 15 | REEnt_10 | 2/8 | 0.13338 | 0.04226 | 0.2406 |
| 15 | REEnt_22 | 2/8 | 0.13338 | 0.04226 | 0.2406 |
| 16 | REEnt_14 | 1/8 | 0.13586 | 0.03341 | 0.19736 |

**Table 3.** T-test for the values of closeness coefficient (CC) in Renewable Energy small medium enterprises (SMEs) in Thessaloniki Prefecture.

|  | Levene's Test for Equality of Variances | t-Test for Equality of Means | |
|---|---|---|---|
|  | Sig. | Sig. (2-Tailed) | Mean Difference |
| Equal variances assumed | 0.642 | 0.028 | −0.319 |
| Equal variances not assumed | 0.642 | 0.028 | −0.319 |

### 3.2. Benchmarking the Sustainable Renewable Energy SMEs

According to the results of the research, the Renewable Energy SMEs achieve a good level of sustainability but not the optimum, as none of them achieve all the characteristics of sustainability. However, the entrepreneurs that are interested in integrating sustainable development at their enterprise level should integrate the main characteristics (Figure 2):

- modern environmental policy
- sustainable marketing
- green network framework
- certified environmental management system

While environmental management standards and frameworks provide a variety of effective tools for bringing significant improvements to the environmental performance of the enterprises, they are limited on developing environmental policies, strategies and procedures [68,69]. Hansmann and Claudia [70] highlighted that the fact that an entrepreneur can successfully address the environmental challenges, it indicates that he can successfully create competitive advantages to add massive value to his products or services [71]. For example, Arnold and Hockerts [72] studied the corporate sustainability innovation strategy of Royal Philips and present some of the sustainability-oriented approaches in the enterprise such as the provision of information regarding sustainability issues, the adoption of sustainability and integrating reporting, the stake-holder integration in environmental issues and the ISO 14001 certification of all parties [73].

**Figure 2.** Benchmarking the sustainable Renewable Energy SMEs.

## 4. Discussion

Environmental sustainability constitutes an important goal in public policies as natural resources are continually eliminated and for this reason governments implement strategic management in order to keep their use within sustainable limits [74,75]. Within this framework, the objectives of corporate sustainability are both socio-economic and environmental and although they may appear to be independent by their nature, they are "inextricably connected and internally interdependent" [76,77]. Recognizing the need to achieve sustainable development, various favourable policies and strategies for the renewable energy industry have been developed, funding has increased and therefore, the enterprises that promote renewable energy, which are continually mushrooming [78]. In order to address the challenges of sustainability, renewable energy resources and environmental strategy, entrepreneurs should evaluate their technological needs and develop a creative strategy in their marketing plan in order to remain competitive and keep a steady pace with the growing sophistication of eco-friendly products and services [79].

The Renewable Energy SMEs performance in the Internet in Thessaloniki Prefecture were studied and analysed regarding their characteristics of sustainability using TOPSIS method. As for the fulfilment of the characteristics of sustainability, most of the Renewable Energy SMEs provide information on their green services and activities and thematic tabs about eco-friendly tips and their green success stories. Also, most of them develop green shopping policy. CC presents values between 0.19736 and 0.66281 and that indicates a great difference between the first and the last enterprise in the ranking. The Renewable Energy SMEs that present "high sustainability" can be used as benchmarks for the enterprises that have been characterized by "low sustainability." The Renewable Energy SME with the best CC makes environmental policy, develops green shopping policy, provides information on green services, describes some green success stories, implements certified environmental management system and makes a review of important environmental impacts.

According to the results of the research, the Renewable Energy SMEs achieve a good level of sustainability through their Internet performance but not the optimum. Chang and Cheng [80] confirm that unlike large enterprises, small and medium-sized enterprises have considerable difficulty in achieving sustainable enterprises. It has to be mentioned that in some cases, the choice of market may constitute the main explanation for differences in sustainable development rate at enterprise rate [81].

However, the entrepreneurs should adopt modern environmental policy, sustainable marketing, green network framework and certified environmental management system in order to consider their enterprise sustainable. Caldera et al. [82] also include risk profiling and ongoing education and awareness in the sustainable business characteristics that enable the enterprises to identify performance improvement opportunities for sustainability transformation. Hao et al. [83] highlight that high-level managerial skills are essential for the entrepreneurs to developing path and practices towards sustainable entrepreneurship internally and also high-level technical skills are essential to integrating new emerging technologies and sustainable characteristics externally.

This research provides the entrepreneurs with an overview on the level of sustainability of the Renewable Energy SMEs. The results can be an efficient tool for entrepreneurs while enhancing the profile of their sustainable enterprise. Undoubtedly, the level of sustainability of an enterprise constitutes a significant characteristic for the awakened customers. Although the paper does not study in detail any particular characteristic, it constitutes a good starting point for entrepreneurs in this sector to get familiarized with the most frequently implemented sustainability management tools. However, the findings provide an overview of the current situation in the second-largest city in Greece, which makes the results less generalizable. So, a future extend in this process would be to search the Renewable Energy Enterprises in the Internet located in the rest of Greece, proceed with a comparison study and conclude to a sustainable benchmark as a tool for continuous sustainable improvement. Finally, some other characteristics could be studied such as life cycle assessment, environmental accounting, organic labels, ecomapping and so forth. that influence the sustainable performance of an enterprise, too.

**Author Contributions:** Z.A. conceived the presented idea of the paper. C.K. developed the theory and performed the computations. Z.A. verified the analytical methods and supervised the findings of the work. Both of them discussed the results and contributed to the final version of the manuscript.

**Funding:** This research received no external funding.

**Conflicts of Interest:** The authors declare no conflict of interest.

## Appendix A

Table A1. Internet research results.

|  | $X_1$ | $X_2$ | $X_3$ | $X_4$ | $X_5$ | $X_6$ | $X_7$ | $X_8$ |
|---|---|---|---|---|---|---|---|---|
| REEnt_1 | 0 | 0 | 1 | 1 | 0 | 1 | 1 | 0 |
| REEnt_2 | 0 | 1 | 1 | 1 | 1 | 0 | 0 | 0 |
| REEnt_3 | 0 | 0 | 1 | 1 | 0 | 1 | 1 | 0 |
| REEnt_4 | 0 | 0 | 1 | 1 | 0 | 0 | 0 | 0 |
| REEnt_5 | 0 | 0 | 1 | 1 | 0 | 0 | 0 | 0 |
| REEnt_6 | 1 | 0 | 1 | 1 | 0 | 1 | 1 | 1 |
| REEnt_7 | 0 | 1 | 1 | 1 | 1 | 0 | 1 | 0 |
| REEnt_8 | 0 | 0 | 0 | 1 | 0 | 1 | 0 | 0 |
| REEnt_9 | 0 | 1 | 1 | 1 | 1 | 1 | 0 | 1 |
| REEnt_10 | 0 | 0 | 1 | 1 | 0 | 0 | 0 | 0 |
| REEnt_11 | 0 | 1 | 1 | 1 | 1 | 0 | 0 | 0 |
| REEnt_12 | 0 | 1 | 1 | 0 | 0 | 0 | 0 | 0 |
| REEnt_13 | 0 | 1 | 0 | 1 | 1 | 1 | 1 | 0 |
| REEnt_14 | 0 | 0 | 0 | 0 | 0 | 1 | 0 | 0 |
| REEnt_15 | 0 | 0 | 1 | 1 | 0 | 1 | 0 | 0 |
| REEnt_16 | 0 | 1 | 0 | 1 | 0 | 0 | 0 | 0 |
| REEnt_17 | 0 | 1 | 0 | 1 | 0 | 1 | 0 | 0 |
| REEnt_18 | 0 | 1 | 0 | 1 | 0 | 1 | 0 | 0 |
| REEnt_19 | 0 | 1 | 0 | 1 | 1 | 1 | 1 | 1 |
| REEnt_20 | 0 | 0 | 1 | 1 | 0 | 1 | 1 | 0 |
| REEnt_21 | 1 | 1 | 0 | 1 | 0 | 1 | 0 | 0 |
| REEnt_22 | 0 | 0 | 1 | 1 | 0 | 0 | 0 | 0 |
| REEnt_23 | 1 | 0 | 1 | 1 | 0 | 1 | 0 | 0 |

## References

1. Shen, K.Y.; Tzeng, G.H. Advances in Multiple Criteria Decision Making for Sustainability: Modeling and Applications. *Sustainability* **2018**, *10*, 1600. [CrossRef]
2. Gibbons, M. Science's new social contract with society. *Nature* **1999**, *C81*, 402. [CrossRef] [PubMed]
3. Cesaretti, G.P. Il ruolo della green society per il superamento della sfida energetica. *Rivista di studi sulla Sostenibilità* **2012**, *2*, 9–12.
4. Andreopoulou, Z.; Misso, R.; Cesaretti, G.P. Using the Internet to support green business for rural development and environmental protection. *J. Environ. Prot. Ecol.* **2014**, *15*, 726–732.
5. Segura-Salazar, J.; Tavares, L.M. Sustainability in the Minerals Industry: Seeking a Consensus on Its Meaning. *Sustainability* **2018**, *10*, 1429. [CrossRef]
6. Reid, R.S.; Nkedianye, D.; Said, M.Y.; Kaelo, D.; Neselle, M.; Makui, O.; Onetu, L.; Kiruswa, S.; Kamuaro, N.O.; Kristjanson, P.; et al. Evolution of models to support community and policy action with science: Balancing pastoral livelihoods and wildlife conservation in savannas of East Africa. *Proc. Natl. Acad. Sci. USA* **2016**, *113*, 4579–4584. [CrossRef] [PubMed]
7. Siarta, S.; Blochb, R.; Knierima, A.; Bachingerb, J. Development of Agricultural Innovations in Organic Agriculture to Adapt to Climate Change—Results from a Transdisciplinary R&D Project in North-eastern Germany. In Proceedings of the 10th European IFSA Symposium, International Farming Systems Association, Aarhus, Denmark, 1–4 July 2012.
8. Stokols, D.; Fuqua, J.; Gress, J.; Harvey, R.; Phillips, K.; Baezconde-Garbanati, L.; Unger, J.; Palmer, P.; Clark, M.A.; Colby, S.M.; et al. Evaluating transdisciplinary science. *Nicot. Tob. Res.* **2003**, *5*, S21–S39. [CrossRef] [PubMed]
9. Pohl, C. What is progress in transdisciplinary research? *Futures* **2011**, *43*, 618–626. [CrossRef]
10. Siebenhüner, B.; Romina, R.; Franz, E. Social Learning Research in Ecological Economics: A Survey. *Environ. Sci. Policy* **2016**, *55*, 116–126. [CrossRef]
11. Godemann, J. Knowledge Integration: A Key Challenge for Transdisciplinary Cooperation. *Environ. Educ. Res.* **2008**, *6*, 625–641. [CrossRef]
12. Nyang'au, I.; Kelboro, G.; Hornidge, A.K.; Midega, C.; Borgemeister, C. Transdisciplinary Research: Collaborative Leadership and Empowerment towards Sustainability of Push–Pull Technology. *Sustainability* **2018**, *10*, 2378.
13. Pintér, L.; Hardi, P.; Martinuzzi, A.; Hall, J. Bellagio STAMP: Principles for sustainability assessment and measurement. In *Routledge Handbook of Sustainability Indicators*; Routledge: Abingdon-on-Thames, UK, 2018; pp. 51–71.
14. Barnett, M.L.; Henriques, I.; Husted Corregan, B. Governing the Void between Stakeholder Management and Sustainability. *Adv. Strateg. Manag.* **2018**, in press.
15. Lyon, T.P.; Delmas, M.A.; Maxwell, J.W.; Bansal, P.; Chiroleu-Assouline, M.; Crifo, P.; Toffel, M. CSR Needs CPR: Corporate Sustainability and Politics. *Calif. Manag. Rev.* **2018**. [CrossRef]
16. Epstein, M.J. *Making Sustainability Work: Best Practices in Managing and Measuring Corporate Social, Environmental and Economic Impacts*; Routledge: Abingdon-on-Thames, UK, 2018.
17. Misso, R.; Andreopoulou, Z.; Cesaretti, G.P.; Hanna, S.S.; Tzoulis, I. Sustainable development and green tourism: New practices for excellence in the digital era. *J. Int. Bus. Entrep. Dev.* **2018**, *11*, 65–74. [CrossRef]
18. Runhaar, H.C.; Tigchelaar, C.; Vermeulen, W.J.V. Environmental leaders: Making a difference. A typology of environmental leaders and recommendations for a differentiated policy approach. *Bus. Strateg. Environ.* **2008**, *17*, 160–178. [CrossRef]
19. Wiesner, R.; Chadee, D.; Best, P. Managing change toward environmental sustainability: A conceptual model in small and medium enterprises. *Organ. Environ.* **2018**, *31*, 152–177. [CrossRef]
20. Poskrobko, B. Teoretyczne aspekty ekorozwoju. *Ekonomia i Środowisko* **1997**, *1*, 7–20. (In Polish)
21. Bombiak, E.; Marciniuk-Kluska, A. Green Human Resource Management as a Tool for the Sustainable Development of Enterprises: Polish Young Company Experience. *Sustainability* **2018**, *10*, 1739. [CrossRef]
22. Schaltegger, S.; Hansen, E.G.; Lüdeke-Freund, F. Business models for sustainability origins, present research, and future avenues. *Org. Environ.* **2016**, *29*, 3–10. [CrossRef]
23. Waldron, T.L.; Fisher, G.; Pfarrer, M. How social entrepreneurs facilitate the adoption of new industry practices. *J. Manag. Stud.* **2016**, *53*, 821–845. [CrossRef]

24. York, J.; O'Neil, I.; Sarasvathy, S. Exploring environmental entrepreneurship: Identity coupling, venture goals, and stakeholder incentives. *J. Manag. Stud.* **2016**, *53*, 695–737. [CrossRef]
25. Dodd, T.; Orlitzky, M.; Nelson, T. What stalls a renewable energy industry? Industry outlook of the aviation biofuels industry in Australia, Germany, and the USA. *Energy Policy* **2018**, *123*, 92–103. [CrossRef]
26. Miller, K.; Neubauer, A.; Varma, A.; Willians, E. *First Assessment of the Environmental Assistance Programme for SMEs (ECAP)*; DG Environmental and Climate Action: London, UK, 2011.
27. Aguado, E.; Holl, A. Differences of corporate environmental responsibility in small and medium enterprises: Spain and Norway. *Sustainability* **2018**, *10*, 1877. [CrossRef]
28. Kuik, O.; Branger, F.; Quirion, P. Competitive advantage in the renewable energy industry: Evidence from a gravity model. *Renew. Energy* **2019**, *131*, 472–481. [CrossRef]
29. Skrypnyk, A.V.; Tkachuk, V.A.; Andruschenko, V.M.; Bukin, E. Sustainable development facets: Farmland market demand estimation. *J. Secur. Sustain. Issues* **2018**, *7*, 513–525. [CrossRef]
30. Andersson, E.; Arfwidsson, O.; Thollander, P. Benchmarking energy performance of industrial small and medium-sized enterprises using an energy efficiency index: Results based on an energy audit policy program. *J. Clean. Prod.* **2018**, *182*, 883–895. [CrossRef]
31. Jiang, Y.; Xue, X.; Xue, W. Proactive Corporate Environmental Responsibility and Financial Performance: Evidence from Chinese Energy Enterprises. *Sustainability* **2018**, *10*, 964. [CrossRef]
32. Bloomberg. *Global Trends in Renewable Energy Investment 2016*; Bloomberg New Energy Finance: London, UK, 2016.
33. Zeng, S.; Jiang, C.; Ma, C.; Su, B. Investment efficiency of the new energy industry in China. *Energy Econ.* **2018**, *70*, 536–544. [CrossRef]
34. Maksimovic, M. Greening the future: Green internet of things (G-IoT) as a key technological enabler of sustainable development. In *Internet of Things and Big Data Analytics toward Next-Generation Intelligence*; Springer: Cham, Switzerland, 2018; pp. 283–313.
35. Aryanto, V.D.W.; Wismantoro, Y.; Widyatmoko, K. Implementing Eco-Innovation by Utilizing the Internet to Enhance Firm's Marketing Performance: Study of Green Batik Small and Medium Enterprises in Indonesia. *IJEBR* **2018**, *14*, 21–36. [CrossRef]
36. Andreopoulou, Z. Green Informatics: ICT for green and Sustainability. *Agrárinformatika* **2012**, *3*, 1–8. [CrossRef]
37. Andreopoulou, Z. Design and Implementation of Model Website for the Promotion of GOingREEN Project for Greener Enterprises in Campania Region. In *GOingREEN. A Collaborative Platform for the Excellences of Campania Region: A Collaborative Platform for the Excellences of Campania Region*; Franco Angeli s.r.l.: Milano, Italy, 2015.
38. Koliouska, C.; Andreopoulou, Z.; Misso, R.; Borelli, I.P. Regional sustainability: National forest parks in Greece. *IJAEIS* **2017**, *8*, 29–40. [CrossRef]
39. Global Industry Analysts. *Green Marketing: A Global Strategic Business Report*; Global Industry Analysts: San Jose, CA, USA, 2012.
40. Groening, C.; Sarkis, J.; Zhu, Q. Green marketing consumer-level theory review: A compendium of applied theories and further research directions. *J. Clean. Prod.* **2018**, *172*, 1848–1866. [CrossRef]
41. Chong, W.K.; Man, K.L.; Kim, M. The impact of e-marketing orientation on performance in Asian SMEs: A B2B perspective. *Enterp. Inf. Syst.* **2018**, *12*, 4–18. [CrossRef]
42. Chan, E.S.W. Gap analysis of green hotel marketing. *Int. J. Contemp. Hosp. Manag.* **2013**, *25*, 1017–1048. [CrossRef]
43. Ardito, L.; D'Adda, D.; Messeni Petruzzelli, A. Mapping innovation dynamics in the Internet of Things domain: Evidence from patent analysis. *Technol. Forecast. Soc. Chang.* **2017**, *218*, 317–330. [CrossRef]
44. Mirghafoori, S.H.; Andalib, D.; Keshavarz, P. Developing Green Performance through Supply Chain Agility in Manufacturing Industry: A Case Study Approach. *Corp. Soc. Responsib. Environ. Manag.* **2017**, *24*, 368–381. [CrossRef]
45. Ardito, L.; Dangelico Rosa, M. Firm Environmental Performance under Scrutiny: The Role of Strategic and Organizational Orientations. *Corp. Soc. Responsib. Environ. Manag.* **2018**, *25*, 426–440. [CrossRef]
46. Niesten, E.; Jolink, A.; Jabbour, A.; Chappin, M.; Lozano, R. Sustainable collaboration: The impact of governance and institutions on sustainable performance. *J. Clean. Prod.* **2017**, *155*, 1–6. [CrossRef]

47. Severo, E.A.; de Guimaraes, J.C.F.; Dorion, E.C.H. Cleaner production and environmental management as sustainable product innovation antecedents: A survey in Brazilian industries. *J. Clean. Prod.* **2017**, *142*, 87–97. [CrossRef]
48. Kernel, P. Creating and implementing a model for sustainable development in tourism enterprises. *J. Clean. Prod.* **2005**, *13*, 151–164. [CrossRef]
49. Liao, Z. Environmental policy instruments, environmental innovation and the reputation of enterprises. *J. Clean. Prod.* **2018**, *171*, 1111–1117. [CrossRef]
50. Milutinović, B.; Stefanović, G.; Dassisti, M.; Marković, D.; Vučković, G. Multi-criteria analysis as a tool for sustainability assessment of a waste management model. *Energy* **2014**, *74*, 190–201. [CrossRef]
51. Hwang, C.L.; Yoon, K. Methods for multiple attribute decision making. In *Multiple Attribute Decision Making*; Springer: Berlin/Heidelberg, Germany, 1981; pp. 58–191.
52. Wang, T.; Liu, J.; Li, J.; Niu, C. An integrating OWA–TOPSIS framework in intuitionistic fuzzy settings for multiple attribute decision making. *Comput. Ind. Eng.* **2016**, *98*, 185–194. [CrossRef]
53. Belenson, S.M.; Kapur, K.C. An algorithm for solving multicriterion linear programming problems with examples. *J. Oper. Res. Soc.* **1973**, *24*, 65–77. [CrossRef]
54. Zeleny, M. A concept of compromise solutions and the method of the displaced ideal. *Comput. Oper. Res.* **1974**, *1*, 479–496. [CrossRef]
55. Yue, Z. TOPSIS-based group decision-making methodology in intuitionistic fuzzy setting. *Inf. Sci.* **2014**, *277*, 141–153. [CrossRef]
56. Park, J.H.; Park, I.Y.; Kwun, Y.C.; Tan, X. Extension of the TOPSIS method for decision making problems under interval-valued intuitionistic fuzzy environment. *Appl. Math. Model.* **2011**, *35*, 2544–2556. [CrossRef]
57. Jahanshahloo, G.R.; Lotfi, F.H.; Izadikhah, M. An algorithmic method to extend TOPSIS for decision-making problems with interval data. *Appl. Math. Comput.* **2006**, *175*, 1375–1384. [CrossRef]
58. Chen, T.Y. The inclusion-based TOPSIS method with interval-valued intuitionistic fuzzy sets for multiple criteria group decision-making. *Appl. Soft Comput.* **2015**, *26*, 57–73. [CrossRef]
59. Joshi, D.; Kumar, S. Interval-valued intuitionistic hesitant fuzzy Choquet integral based TOPSIS method for multi-criteria group decision making. *Eur. J. Oper. Res.* **2016**, *248*, 183–191. [CrossRef]
60. Sadi-Nezhad, S.; Damghani, K.K. Application of a fuzzy TOPSIS method base on modified preference ratio and fuzzy distance measurement in assessment of traffic police centers performance. *Appl. Soft Comput.* **2010**, *10*, 1028–1039. [CrossRef]
61. Kelemenis, A.; Askounis, D. A new TOPSIS-based multi-criteria approach to personnel selection. *Expert Syst. Appl.* **2010**, *37*, 4999–5008. [CrossRef]
62. Chen, S.J.; Hwang, C.L. *Fuzzy Multiple Attribute Decision Making: Methods and Applications*; Springer: Berlin, Germany, 1992.
63. Yoon, K.P.; Hwang, C.L. *Multiple Attribute Decision Making*; Sage Publication: Thousand Oaks, CA, USA, 1995.
64. Behzadian, M.; Otaghsara, S.K.; Yazdani, M.; Ignatius, J. A state-of the-art survey of TOPSIS applications. *Expert Syst. Appl.* **2012**, *39*, 13051–13069. [CrossRef]
65. Hatami-Marbini, A.; Kangi, F. An extension of fuzzy TOPSIS for a group decision making with an application to Tehran stock exchange. *Appl. Soft Comput.* **2017**, *52*, 1084–1097. [CrossRef]
66. Zeleny, M.; Cochrane, J.L. *Multiple Criteria Decision Making*; University of South Carolina Press: Columbia, SC, USA, 1973.
67. García-Cascales, M.S.; Lamata, M.T. On rank reversal and TOPSIS method. *Math. Comput. Model.* **2012**, *56*, 123–132. [CrossRef]
68. Curkovic, S.; Sroufe, R.; Melnyk, S. Identifying the factors which affect the decision to attain ISO 14000. *Energy* **2005**, *30*, 1387–1407. [CrossRef]
69. Eltayeb, T.K.; Zailani, S.; Ramayah, T. Green supply chain initiatives among certified companies in Malaysia and environmental sustainability: Investigating the outcomes. *Resour. Conserv. Recycl.* **2011**, *55*, 495–506. [CrossRef]
70. Hansmann, K.W.; Claudia, K.; Environmental Management Policies. *Green Manufacturing and Operations: From Design to Delivery and Back*; Routledge: Abingdon-on-Thames, UK, 2001; pp. 192–204.
71. Mishra, D.; Gunasekaran, A.; Papadopoulos, T.; Hazen, B. Green supply chain performance measures: A review and bibliometric analysis. *Sustain. Prod. Consum.* **2017**, *10*, 85–99. [CrossRef]

72. Arnold, M.G.; Hockerts, K. The greening dutchman: Philips' process of green flagging to drive sustainable innovations. *Bus. Strateg. Environ.* **2011**, *20*, 394–407. [CrossRef]
73. Jayaram, J.; Avittathur, B. Green supply chains: A perspective from an emerging economy. *Int. J. Prod. Econ.* **2015**, *164*, 234–244. [CrossRef]
74. Andreopoulou, Z.S.; Manos, B.; Polman, N.; Viaggi, D. *Agricultural and Environmental Informatics, Governance, and Management: Emerging Research Applications*; IGI Global: Hershey, PA, USA, 2011.
75. Andreopoulou, Z.; Koliouska, C.; Galariotis, E.; Zopounidis, C. Renewable energy sources: Using PROMETHEE II for ranking websites to support market opportunities. *Technol. Forecast. Soc. Chang.* **2018**, *131*, 31–37. [CrossRef]
76. Bansal, P. The corporate challenges of sustainable development. *Acad. Manag. Exec.* **2002**, *16*, 122–131. [CrossRef]
77. Hahn, T.; Figge, F.; Pinkse, J.; Preuss, L. A paradox perspective on corporate sustainability: Descriptive, instrumental, and normative aspects. *J. Bus. Ethics* **2018**, *148*, 235–248. [CrossRef]
78. Wang, P.; Lu, Z.; Sun, J. Influential Effects of Intrinsic-Extrinsic Incentive Factors on Management Performance in New Energy Enterprises. *Int. J. Environ. Res. Public Health* **2018**, *15*, 292. [CrossRef] [PubMed]
79. Tesar, G.; Moini, H.; Sorensen, O.J. *Mapping Managerial Implications of Green Strategy: A Framework for Sustainable Innovation*; World Scientific: Hackensack, NJ, USA, 2018.
80. Chang, A.-Y.; Cheng, Y.-T. Analysis model of the sustainability development of manufacturing small and medium-sized enterprises in Taiwan. *J. Clean. Prod.* **2019**, *207*, 458–473. [CrossRef]
81. O'Gorman, C. The sustainability of growth in small-and medium-sized enterprises. *Int. J. Entrep. Behav. Res.* **2001**, *7*, 60–75. [CrossRef]
82. Caldera, H.T.S.; Desha, C.; Dawes, L. Exploring the characteristics of sustainable business practice in small and medium-sized enterprises: Experiences from the Australian manufacturing industry. *J. Clean. Prod.* **2018**, *177*, 338–349. [CrossRef]
83. Hao, Y.; Helo, P.; Shamsuzzoha, A. Virtual factories for sustainable business performance through enterprise portal. *Int. J. Comput. Integr. Manuf.* **2018**, *31*, 562–578. [CrossRef]

© 2018 by the authors. Licensee MDPI, Basel, Switzerland. This article is an open access article distributed under the terms and conditions of the Creative Commons Attribution (CC BY) license (http://creativecommons.org/licenses/by/4.0/).

Article

# Environmental Protection in School Curricula: Polish Context

Anna Mróz [1,*], Iwona Ocetkiewicz [1] and Katarzyna Walotek-Ściańska [2]

1. Faculty of Pedagogy, Pedagogical University of Cracow, Podchorążych 2, 30-084 Kraków, Poland; iwona.ocetkiewicz@up.krakow.pl
2. Faculty of Philosophy, Jesuit University Ignatianum in Krakow, Mikołaja Kopernika 26, 31-501 Kraków, Poland; katarzynaws@interia.pl
* Correspondence: anna.mroz@up.krakow.pl

Received: 30 September 2018; Accepted: 29 November 2018; Published: 3 December 2018

**Abstract:** Properly planned and effectively implemented education provides an opportunity to change human behavior, which in turn may lead to an improved quality of life worldwide, including by means of realizing a cleaner environment. This article presents the results of research on the integration of environmental protection issues into curricula by Polish teachers. It was assumed that the environmental protection issues included the challenges related to the sustainable management of natural resources. The sample consisted of 337 teachers of general subjects who were employed in schools in the Małopolska region (southern Poland) and working with students in lower-secondary (13–16 years old) and upper-secondary (16–20 years old) schools. The results of the research show that many teachers know how to integrate environmental protection issues into their curricula. However, there are still many teachers who ignore key issues in the education of sustainable development in their teaching process.

**Keywords:** education for sustainable development; environmental protection; curriculum; teacher; renewable resources

## 1. Introduction

### 1.1. EnvironmentalProtection in the Educational Context

The second half of the 20th century and beginning of the 21st is undoubtedly a time of great ecological crisis, the results of which may be disastrous for present and future generations. Uncontrolled globalization and the introduction of new technologies, many of which have forever changed the methods of manufacturing basic goods and services, have resulted in the destruction of ecosystems and the degradation of the natural environment, threatening life on Earth. The massive consumption of fossil fuels has caused visible damage to the environment in various forms. Approximately 90% of the energy we consume comes from fossil fuels [1]. One of the causes of this dangerous ecological crisis is the overexploitation of natural resources and the neglect of the needs of the natural environment. This is why perhaps the most important issue of our time is how to sustain our planet's resources while still developing the wealth and well-being of a growing population [2].

Often, the behaviors which lead to environmental damage are the result of human ignorance which, in turn, is a consequence of the lack of well-planned and effective education in the area of environmental protection (involving, first of all, the responsible use of natural resources and sustainable management of renewable energy resources). An attempt to compensate for this is a worldwide sustainable development concept which originates from concerns about the future world and the desire for better conditions for future generations. As an idea, sustainable development had already been espoused in the World Conservation Strategy (WCS), a document produced in 1980 by the International Union for the Conservation of Nature, World Wildlife Fund (now the Worldwide Fund for Nature) and the United Nations Environmental Programme [3]. The idea of sustainable development is promoted and supported by *education for sustainable development*, also called *education for sustainability*, *environmental education*, and *sustainability education*. Implementation of its principles in a didactic process should, first of all, ensure the harmonious (understood as undelayed, applied correctly and at the right time in all spheres of development) development of students and enable them to face new challenges in the future. According to the sustainable development paradigm, a well-designed education process shapes key competencies which will allow students to develop harmoniously, function actively in the present, as well as make responsible decisions and support the sustainable development of society in the future.

Education is an important element of the enhancement of pupils' environmental awareness. This is because education plays an important role in shaping and transforming society [4]. The concept of *education for sustainable development* (ESD) was outlined in the Agenda 21 document created during the Earth Summit in 1992, section IV—titled *Means of Implementation*—in chapter 36: *Promoting education, public awareness and training in the area of balanced and sustainable development and environmental protection*. In this document, education is considered the key factor in the implementation of sustainable development principles. Since then, there have been many documents and publications addressing the principles of education for sustainable development, including its objectives, the means of implementation, and recommended methods and forms of work.

It states in the Agenda and in documents related to the Decade of Education for Sustainable Development that education, raising public awareness, and additional training of people around the world—by means of formal, informal, and non-formal methods—are critical areas of activity for sustainable development. Suitable educational initiatives and the introduction of the new education model to schools will enable the sustained and harmonized development of humanity and, in consequence, the world. Additionally, it was declared that education worldwide must shift toward the issues of balanced and sustainable development and environmental protection [5]. It was emphasized that education—institutionalized education, in particular—should be recognized as a process by which human beings and societies can reach their fullest potential. It should also be integrated into all disciplines by employing different didactic methods and effective means of communication.

In 2015, a new plan for the implementation of sustainable development principles was developed—Agenda 2030—which set 17 Sustainable Development Goals to achieve and 169 targets to meet by 2030 on a global scale. These objectives address five areas, the so-called 5xP: people, planet, prosperity, peace, and partnership. The goals cover a wide range of challenges, such as poverty, hunger, health, education, gender equality, climate changes, sustainable development, peace, and social justice [6]. They succeeded the Millennium Development Goals, which were to be achieved by 2015.

Education for sustainable development is commonly understood as education that encourages changes in knowledge, skills, values, and attitudes to enable a more sustainable and just society for all. ESD aims to empower and equip present and future generations to meet their needs using a balanced and integrated approach to the economic, social, and environmental dimensions of sustainable development [7].

Education for a sustainable future is a huge challenge for education systems. It requires the following questions to be addressed: How can we better understand the complexity of our world? How are the problems of today's world interrelated, and what do they mean to the people who try to solve them? What kind of world do we want in the future? Does this vision fit within the systems sustaining life on Earth? How can we reconcile the requirements of the economy, societies, and the environment? The idea of ESD is the commitment to achieving a balance between social and economic well-being, and between culture and tradition and the protection of natural resources. ESD emphasizes the need to respect human dignity, honor diversity, and protect the natural environment and natural resources on our planet [8–11].

The scope and diversity of current environmental issues have made it necessary to define several problem areas—such as the use of natural resources, pollution, health and nutrition, urban areas, etc.—while paying due attention to geographical, ecological, and economic variations [10]. Investigating knowledge, perceptions, as well as attitudes of the public toward various aspects of environmental issues is of great importance to the promotion of sustainable development [12–14]. It is worth emphasizing that these issues have long been present in environmental education (EE). However, only since the 1970s has there been a shift toward a more holistic approach to EE content and the target group. Now, EE aims to create a population that not only is knowledgeable but also shows positive attitudes and takes action to preserve the environment. According to ESD principles, it is not only ecological education that should address important environmental issues. These problems should be discussed during the teaching of all subjects and at every stage of child and youth education.

The term "environmental protection" includes, among others, the postulate of the sustainable management of natural resources.

Sir Ken Robinson noticed that curricula set out frameworks of what students should know, understand, and be competent to do. In most schools, some parts of the program are obligatory, others are optional, and some are voluntary, like extracurricular activities. Curricula can be formal and informal. A formal program is obligatory, with grades and exams, while an informal component includes voluntary activities [15]. Teachers, then, have much freedom in selecting non-obligatory, voluntary content. They can also, to some extent, choose the content and issues for the obligatory parts (for example, the subjects of home assignments, presentations, or readings during foreign language classes). Curricula allow for addressing certain issues in many ways. The author of the Second Collection of Good Practices Education for Sustainable Development UNESCO Associated Schools 2009 [16] emphasizes that Ministries of Education worldwide examine how to introduce and reinforce key sustainability issues throughout the curricula, in the training of teachers, in extracurricular activities, and in non-formal education (p. 5). It is a very important issue—from the point of view of educational practices but also globally—as it may influence the future condition of our world. The Global Action Programme on Education for Sustainable Development also underscores the integration of key sustainable development issues in school curricula [17].

*1.2. The Teacher's Role in the Challenges of Modern Education*

Ongoing social changes have impacted the vision of the professional role of teachers. The desired professional roles of a teacher working with adolescents and implementing sustainable development education principles are those of a guide, an interpreter, a researcher, a reflective practitioner, an emancipated teacher, a transformative intellectualist, and a post-positivist practitioner [18].

Teachers as guides fulfill their role by showing students new areas of knowledge and setting directions toward which they will lead the students. They are also translators of a complex, diverse, and multi-voice culture; they explain the sources, meanings, and potential consequences of different phenomena. They explain the possible choices that arise in individual and unique ways that advance identity development and suggest ways of dealing with struggles in a multicultural world [18].

As researchers, teachers know that educational investigations should be an integral part of their work. As stated by Stanisław Palka, the "teaching profession forces to assume the role of pedagogical researcher, as teachers must diagnose, explain and make decisions based on, among others, the results of their conversations with students, other teachers and parents, document analysis and analysis of class test and didactic test results" [19]. While studying their own activities, teachers should focus on the teaching process as it connects with discovery and exploration, as well as with the facilitation and improvement of the cognitive activity of students. In this way, teachers may become participants, creators, and competent observers of educational events.

Teachers in the role of reflective practitioners build their competencies based on their own practical experience. The idea of reflective practice originated from the studies by an American epistemologist, Donald Schön, who analyzed competencies of professionals in the U.S. during atime of decreasing social trust toward this profession. He presented his theory in the book *The Reflective Practitioner: How Professionals Think in Action*, published in 1987 [20]. Reflection-in-action became Schön's main focus. He called itthe "experimenting" that takes place in current situations—problematic and uncertain—which differ from previous ones. The goal of reflection-in-action is to change the situation. According to Schön [20], when reflective practitioners try to understand a given situation, they effect a change, which increases their awareness. It is a sequence: understanding inaction, followed by a "reflective spiral". Spontaneous but also smooth and on-the-spot action results from the knowledge found within oneself. It is knowing-in-action which is hard to articulate, but once finished, it can be subject to reflection. Schön explains that reflective practitioners have the ability to see unfamiliar situations as familiar ones and to act as they have in similar circumstances. He calls it "see-as" and "do-as". When reflecting in action, on action, and about action, teachers engage in the active search for solutions, the aim of which is not only to better understanding theself but also to improve their own work. They must constantly update their professional situations and be researchers oftheir own practice, approaching certain situations as research problems and integrating their thinking and doing into thesearch forsolutions, formulating their own theories independent of scientific ones. In this way, they develop their personal knowledge and own ways of acting. A teacher's reflection is a kind of thinking that accompanies doing in order to transform some elements or the whole situation. Reflective practitioners create their workshop by analyzing their own knowledge. Their professional behavior involves not only reflection in and overbut also about the action. This reflection results from acritical attitude which requires the exploration of ethical, political, and instrumental issues rooted in their daily thinking and professional practice. The value of the concept of are flective teacher practitioner is in the fact that personal experience is treated as the source of knowledge necessary to fulfill professional duties. By trusting their own experiences, teachers create their own practice and theory of this practice. Thanks to the emphasis on the subjective activity of teachers, rights and opportunities to participate in group life result from their reflective actions, as well as the actions of students oriented toward intentional adaptation or change.

Another role of teachers—emancipated teachers—is connected with understanding the typical emancipatory pedagogy of student subjectivity. The implementation of emancipatory pedagogy postulates requires respecting the principle of the subjective nature of pedagogical interaction. A subjective sense of freedom is shaped and developed during the process of learning and doing, and its starting point is within personal experience, tangible opposition, and conflict. Emancipated teachers are engaged in change. They oppose the compulsory solutions and want to search for their own alternative options which could replace the hitherto ones. They show advanced critical thinking and reflective abilities in viewing the world. Emancipated teachers are promoters of change and empowerment of education. In addition, being critical and reflective, they develop tools to modify their own practice; learn how to teach by planning constant improvements according to lifelong education postulate; master the process of discovering and creating knowledge, together with knowing the methodologies of the discipline taught and education sciences; develop the critical approaches, openness, and courage necessary to reorient education and the concept of school.

Critical pedagogy representatives, such as Paulo Freire, Henry Giroux, and Peter McLaren, postulate that teacher should be a transformative intellectual (a term coined by Henry Giroux), that is, an advocate of weaker social groups and translator between humans and the complex social world. The transformative intellectual should facilitate a critical understanding of the world, first of all by showing social inequalities [21–24]. The concept of the teacher as a transformative intellectual is rooted in a radical ideology which is contrary to conservative and liberal ideologies. The conservative ideology focuses on the education of students as future employees who need to acquire standard skills, and the role of the teacher is reduced to transmitting a socially accepted scope of knowledge. In the liberal ideology, personal development is the most important. It can be achieved through personal successes and awards when competing with others, and the teacher's role is to create conditions fostering the development of critical thinking in students. Teachers as transformative intellectuals should be open to the problems of human rights, the rights of minorities and marginalized people, and social inequality and exclusion; they should be sensitive to differences in public and private life and follow the road to execute resistance policy. It is worth emphasizing that these problems have been considered by UNESCO as threats to the sustainable development of the world.

The teacher as a post-positivist practitioner responds to present and future challenges. This image is the result of applying critical, postmodern social theory connected with reflective orientation in the analysis of the teacher's work. The image comprises characteristic behaviors of a postmodern teacher, including:

- practicing reflectiveness based on the ability to investigate one's own practice;
- connecting thinking with social and historical contexts and power aspects;
- improvising, thinking-in-action, responding to unexpected situations;
- modeling a culture of activity among students that allows different opinions and attitudes, prefers discussions, and is open to students' experiences and their way of thinking about the world and representing the world;
- exercising critical social and self-reflection based on the dialogue of subjects of the teaching-learning process;
- engaging in the development of democracy in the class and school;
- displaying an active attitude that breaks the students' tendency to passiveness and passive thinking, connecting knowledge with action;
- possessing knowledge of cultural differences in the community, respecting minority rights, cultural pluralism;
- including the emotional side of contact with students in the teaching process, allowing emotional behaviors like humor, compassion, empathy, and resentment in dialogues with students [18].

The model of the teacher as a post-positivist practitioner of the postmodern age is a postulative model which shows how teachers can be, rather than how they must be. These teachers follow the idea of subjectivity, emancipation, and engagement in matters of the world functioning in a democratic and sustainable way. They reject and oppose being enslaved by power, knowledge, and modernist anti-intellectualism. It is easy to notice that the above-presented models of teachers' roles are connected and overlapping. Their responsibilities are repeated and focus mainly on a reflective attitude and critical thinking [18].

All of this suggests that the primary task of modern teachers is to empower students, so they can actively and creatively transform their reality and are prepared to live valuable lives in the global information society. Teachers are responsible for preparing young people for active participation in creating a better future—their own and the world's. Therefore, the style of work is important: the methods and teaching forms, as well as the content addressed during classes, extra activities, and school trips—in other words, the content teachers include in their curricula. This content may translate into the growth of students' knowledge and their daily behaviors, choices, and attitudes and, through those aspects, into their efforts toward sustainable development in three dimensions: social, environmental, and economic.

Teachers who work with adolescents (13–20 years old) must know the needs, abilities, and limitations of their students. The period of learning in lower- and upper-secondary school is the developmental stage called *early* and *late adolescence*. In early adolescence, young people begin to search and gather new physical, social, and intellectual experiences, so they can then organize and unify into new patterns and scenarios that best prepare them for adult life [25]. During the cognitive development of teenagers, there is a shift to formal operations. Early adolescence is marked by the increasingly more advanced implementation of the rules and laws of certain areas of knowledge, expressed verbally or through symbols. Formal thinking allows for going beyond what cannot be known directly; enables reasoning from the general to the particular—from hypothesis or assumption to conclusion; formulating general conclusions based on detailed facts (scientific-inductive thinking). Formal thinking is also the ability to simultaneously consider many variables (combinatory reasoning); abstract thinking reflects understanding by way of inner thoughts or reflection based on the available knowledge but reaching beyond what is observable (reflective-abstract thinking) [25,26]. Unlike children, teenagers are able to consider different options and possibilities. They can imagine the future consequences of their present actions; thus, they are capable of some form of long-term planning. Another characteristic of formal thinking is the search for solutions to problems in a systematic and methodical way. Thinking based on formal operations ensures compensation for imbalances in cognitive structures and adaptation by using the following logical operations: identity, negation, inverse, and reciprocity—that is, searching for one law (rule, principle) that is absolutely true regardless of the context (so-called logical absolutism). As the foundation of the hypothetico-deductive system, classical logic does not allow contradictions, which means that, during adolescence, the dominating means of resolving mental contradictions is, according to formalized thinking, the rejection of one of the alternatives. However, it is worth noticing that some approaches posit that teenagers are able to think relatively and are convinced that knowledge is relative. Adoption of relativist thinking by young people is probably easier due to their contact with social problems (active fulfillment of different and often contradicting social roles; personal motivation included in the cognitive process, etc.) and due to their received training on humanistic knowledge. In turn, frequent training on formal knowledge, such as mathematics, physics, or foreign languages, may facilitate the development of formal thinking. Thus, cognitive development is dynamic in the areas of activity specific to certain individuals. If a student often engages in solving problems from a certain knowledge area, he or she may expect their knowledge and skills in this area to become more advanced, internally connected, and viewed relatively from many perspectives [25,26]. During adolescence, there are also changes in (1) knowledge, (2) cognitive processes and competencies, and (3) meta cognitive orientations. Knowledge refers to the three types of information structures in long-term memory. They include declarative knowledge based on the "I know that..." scheme, covering the known facts; procedural knowledge ("I know how...") regarding all abilities focused on the goal; and conceptual knowledge ("I know why...")—understanding the declarative and procedural knowledge [25].The cognitive processes of adolescents take on different forms: inductive and deductive reasoning, analogies, decision making, and problem-solving. Other important processes refer to coding or formulating a mental representation of the learning situation, memorizing and storing information in long-term memory, and reproducing—taking information out of the long-term memory. Metacognitive orientations refer to adolescents reflecting the state of their knowledge and evaluation of this knowledge—a young person should be able to determine what they know; if they know that they do not know; and whether they tend to absolutize or relativize. Adolescence is a period of intense development of imagination that can be seen in teenagers' creativity and dreams. They tend to see reality in an original way; easily accept innovations, including technical ones; have spontaneous rationalized ideas which make their daily life easier. Their dreams, regarding both themselves and their environment, often allow them to achieve what, in reality, is out of their reach. Creative work is common among young people and is most often expressed in the form of diaries, original music, song lyrics, or poems. In the area of verbal creativity in asocial forum (e.g., editing school newsletters), girls are more active [26]. Memory also changes during adolescence. Memory development can be connected to four components: capacity, base memory processes, memorizing strategies, and metamemory. The source of these transformations are changes in the adolescent brain. On the one hand, compared to adults, young people show some weaknesses in cognitive processing, including weaknesses in the executive function of the frontal lobe, visual-spatial and verbal memory, and response time (i.e., slower). On the other hand, in early adolescence, there is already growth in the volum of memorized and stored information, especially practical information. This is possibly a result of an increased number of neuronal connections that form during adolescence. Ever more plastic and faster processes of recognizing, coding, filing, and searching lead to the improvement of base memory processes. The effectiveness and frequency of using memory strategies are also enhanced, along with the ability to memorize and improve metamemory [26]. During adolescence, young people need to develop their emotional-cognitive structures, which allow them to succeed in difficult situations and build proper emotional relationships with others. The most important emotional competencies teenagers need to develop are the following: regulation of the intensity of emotions, as they tend to experience extreme emotions; adaptation of the means of expressing rapidly changing (unsteady) emotions; calming down; understanding the social consequences of expressing emotions openly or hiding them; using symbolic thinking to transform negative emotions into less repulsive ones; discerning emotions from the facts that cause them; maintaining relationships with other people despite strong negative emotions; acquiring a wide range of skills in experiencing and expressing empathy; and using cognitive skills to better understand the nature and the sources of emotions. Better cognitive abilities and the search for distinctness in *me-others* relationships become the basis of more mature, empathetic experiences. The ability to feel empathy enables young people to feel the suffering and distress of others, even when there is no direct situational pressure. This may motivate engagement in prosocial activities and strengthen their social sensitivity. During middle and late adolescence, some additional higher feelings connected with an upheld value system or ideology are developed. The main values preferred by young people are truth, virtue, religious values, altruism, sociopolitical values underlying tolerance, justice, and others [25,26]. Therefore, adolescence is theperfect time to shape the knowledge and attitudes of young people, including those regarding the needs of the natural environment.

## 2. Methodological Assumptions of the Research

Our analyses focus on the problem of integrating environmental content, including the sustainable management of natural resources, into Polish curricula for students aged 13–20. We assumed that the frequency and means of addressing the critical ESD issue by teachers translate into the knowledge students have about environmental protection, which, in turn, may determine the awareness, attitudes, and, consequently, environmental behaviors of young people. We also assumed that, according to UNESCO experts' postulates, all teachers should include topics critical for ESD into their curricula. As a UN and EU member, Poland must respect the principles set in the Agenda 2030 [26] and its 17 priority goals for a sustainable future. Goal 4 is to *Ensure inclusive and equitable quality education and promote lifelong learning opportunities for all* [27], whereas Goal 7 is to *Ensure access to affordable, reliable, sustainable and modern energy for all*. We assumed that Goal 7 may be achieved through education (Goal 4). A. Leicht, J. Heiss, and W. J. Byun [6] argue that "Educational programmes, particularly non-formal and informal, can promote better energy conservation and uptake of renewable energy sources" (p. 31).

### 2.1. Main Goal of the Study

The main goal of the study isto explore ways of integrating environmental issues into curricula in lower- and upper-secondary schools (students aged 13–20). We assumed that the term *nature protection* means all activities to preserve animate and inanimate features of nature and landscapes in their unchanged or optimal state. The main environmental protection objective is to maintain the stability of ecosystems and ecological processes, as well as biological diversity. One of the elements of nature protection is the promotion of renewable energy sources. There is an unquestionable correlation between generating energy from traditional sources and the degradation of the natural environment. Sustainable development also refers to human activities which maintain the balance of nature and aim toprovide present and future generations with the opportunity to meet their needs and enjoy nature. Thus, it is contrary to this principle to exhaust nonrenewable resources and cause irreversible climate change (and its consequences for the environment) and dangerous air pollution. To summarize, sustainable development (including natural environment protection) in the context of environmental protection means—among other things—actions that promote the use of renewable energy sources.

*2.2. Method, Research Technique, and Tool*

We collected data by means of a survey (survey method) [28]. The study was performed within the quantitative paradigm [29].The use of a survey allows for the collection of answers from many respondents quickly, and respondents' declarations are precise and refer to the exact subject matterwhich is of interest to researchers. The technique used in the research was polling. In order to collect data, we used questionnaires. The tool was developed for the purposes of this particular study (see Table 1). Respondents were asked to indicate which critical sustainable developmenttopics are regularly (not occasionally or from time to time) part of their curricula, and how frequently are they introduced.

Table 1. Method, technique, and research tool used in the research.

| Method | Survey |
|---|---|
| Research technique | polling |
| Tool | original questionnaire, developed for the purposes of this particular research |

On the basis of literature review, we identified four main ways of integrating environmental protection issues into teaching programs:

(1) Covering ESD-related topics during subject lessons;
(2) Covering ESD-related topics during additional classes: for example, interest clubs;
(3) Assigning homework referring to sustainable development topics;
(4) Encouraging students to engage in out-of-school activities.

We determined the dependent and independent variables (Table 2).

Table 2. Variables determined in the research.

| Dependent Variable | Independent Variables |
|---|---|
| integrating environmental protection issues into school curricula | - School level (*gimnazjum*—lower-secondary school; *lyceum/technikum*—upper-secondary school)<br>- School location (village, small town, big city)<br>- Years of employment (0–5, 6–10, 11–15, 16–20, 21 and more)<br>- Degree of professional promotion of teachers (trainee, contractual teacher, appointed teacher, chartered teacher)<br>- Subject taught (Polish language, foreign languages, human and social sciences, science and natural sciences, mathematics, and information and technical subjects). |

Using Pearson's chi-square test, we checked which independent variables are statistically relevant for including environmental issues in subject curricula.

*2.3. Sample and Organization of the Research*

In order to collect the research data, convenience sampling [30] was applied in accordance with the following criterion: the consent of the participants to take part in the survey. The statistical and demographic data of teachers who participated in the survey are presented in Table 3. and Figure 1.

Table 3. Statistical and demographic data of teachers who participated in the survey.

| School location | N | % |
|---|---|---|
| Village | 128 | 37.98 |
| Small town | 136 | 40.36 |
| Big city | 68 | 20.18 |
| N/A | 5 | 1.48 |
| School level | N | % |
| Lower secondary | 177 | 52.52 |
| Upper secondary | 149 | 44.21 |
| N/A | 11 | 3.26 |
| Years of employment | N | % |
| 0–5 years | 36 | 10.68 |
| 6–10 years | 64 | 18.99 |
| 11–15 years | 85 | 25.22 |
| 16–20 years | 102 | 30.10 |
| 21 years and more | 50 | 15.0 |
| Degree of professional promotion | N | % |
| Trainee | 13 | 3.86 |
| Contractual teacher | 45 | 13.35 |
| Appointed teacher | 82 | 24.33 |
| Chartered teacher | 196 | 58.16 |
| Subject taught | N | % |
| Polish language | 60 | 17.80 |
| Human and social sciences (civil knowledge, history, knowledge about culture, entrepreneurship) | 67 | 19.88 |
| Foreign languages | 74 | 21.96 |
| Mathematics | 36 | 10.68 |
| Science and natural sciences (biology, chemistry, physics, geography, education to family life) | 86 | 25.52 |
| Information and technical subjects (ITC, technics) | 9 | 2.67 |
| Physical education | 1 | 0.30 |
| Arts (music, visual arts) | 3 | 0.89 |
| NS | 1 | 0.30 |
| Total | 337 | 100% |

Teachers were asked about the key sustainable development issues that they integrate into their curricula and the waysby which theypresent and discuss these issues. The survey was conducted from May to November 2016. Teachers of general subjects in lower- and upper-secondary schools in the Małopolskie region (*voivodship*) were invited to take part in the research. There were 927 questionnaires distributed, out of which 337 were completed. The survey was conducted in:

- the 2 biggest cities of the Małopolska region (8 lower- and 3 upper-secondary schools);
- 9 small towns (16 lower- and 14 upper-secondary schools);
- 21 villages (19 lower- and 2 upper-secondary schools).

For the purpose of this text, the issues of environmental protection were analyzed. Teachers were instructed that environmental protection means any action designed to remedy or prevent damage to physical surroundings or natural resources by a beneficiary's own activities, to reduce the risk of such damage, or to lead to a more efficient use of natural resources, including energy-saving measures and the use of renewable sources of energy [31].

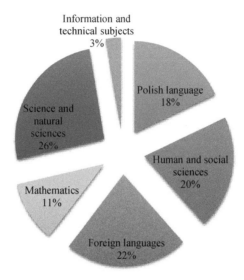

**Figure 1.** Percentage participation of teachers of particular subjects in the research.

The Małopolska region, where the research was carried out, is located in the central-southern part of the country. The region has 3.4 million inhabitants (as of December 2017), making it the fourth most populated region in Poland. The population density is the second highest in the country. In terms of economic development, the region is one of the most diverse in Poland. Situated at the crossing point of communication routes between the West (Austria, Germany) and the East (Ukraine) and between the North (Scandinavian countries) and the South (Slovakia, Hungary), it has exceptional assets and annually attracts 10–12 million tourists, new investments, and young people studying at regional universities. Małopolska has a high potential for scientific research and higher education—Cracow is the second center of research development in Poland. There are 32 higher-education institutions and universities, representing 7.1% of all higher-education institutions in Poland. There are over 180,000 students and more than 55,000 graduates per year. In 2013, over 25% of entrepreneurs' spending on innovations was allocated to research and development (first place in the country). Małopolska is one of the most important Polish bioregions, with a high potential for the development of life sciences with well-developed medical facilities. This region has a high potential for energy technology research. The capital city of Małopolska—Cracow—is one of the most important industrial designer education centers in Poland. In Małopolska, there are more than 100 research and development centers, including the Foundry Research Institute, Institute of Forensic Research, Institute of Oil and Gas, and Institute of Advanced Manufacturing Technologies [32]. Due to the significant air pollution in the biggest city (Cracow), there is an ongoing debate on alternative and sustainable sources of energy. Considering the fact that Małopolska is a region with many universities, communication between scientists and the regional population is easy, and this may translate into a higher level of knowledge and awareness of the issues of environmental protection among the residents. It is a problem that affects all the residents, which, in turn, obligates teachers to search for solutions in their daily practice by, among other actions, integrating this topic into their curricula.

## 3. Research Results

### 3.1. Integrating Environmental Protection Issues into Curricula

Our analysis indicates that teachers of all subjects are aware of the need to cover the issues of environmental protection in their curricula (Table 4). Unsurprisingly, these topics are most often introduced by science and natural science teachers (biology, physics, chemistry). It is interesting that even more foreign language teachers (87.84%) declare they cover environmental protection issues in their curricula. Perhaps, being aware of the principles of ESD, they try to introduce topics critical for the sustainable future of the world. We need to point out that, in Poland, there are many UNESCO resources on ESD, available in English, German, or French. This means that teachers of these subjects have easier access to the materials.

**Table 4.** Integration of environmental protection into curricula by teachers of various subjects.

| Polish Language | Humanities and Social Sciences | Foreign Languages | Mathematics | Natural Sciences | Information and Technical Subjects | Total |
|---|---|---|---|---|---|---|
| % | % | % | % | % | % | % |
| 60.00 | 59.7 | 87.84 | 63.89 | 86.06 | 55.56 | 73.59 |

The study reveals a small difference between the declarations made by lower- and upper-secondary school teachers. Slightly more teachers working in lower-secondary education teach ESD-related topics. This may be due to the fact that lower-secondary school teachers are not pressured by the need to prepare students formatriculation exams and can spend more time on additional topics critical for the future of our world and for ensuring a better quality of life for future generations. The results of analyses of dependencies between education level taught and whether environmental protection issues are addressed during classes are presented in Table 5.

**Table 5.** Integration of environmental protection into curricula and level of education taught.

| Lower Secondary | Upper Secondary | Total |
|---|---|---|
| % | % | % |
| 78.53 | 67.11 | 73.59 |

During the analysis, we noticed the following relation (see Table 6): the more years of employment, the higher the percentage of teachers declaring that they integrate environmental protection issues into their curricula. Therefore, if teachers with more professional experience are more willing to include environmental protection issues into their curricula, this may indicate that teachers with greater professional experience have greater knowledge about the principles of ESD and are more willing to integrate its key topics into their curricula.

**Table 6.** Integration of environmental protection into curricula and years of employment.

| 0–5 | 6–10 | 11–15 | 15–20 | 21 and more | Total |
|---|---|---|---|---|---|
| % | % | % | % | % | % |
| 58.33 | 59.38 | 78.82 | 80.26 | 91.16 | 73.59 |

There are five stages of teacher promotion in Poland: university graduates become "trainees" who teach for 1–2years. After few years of work, a trainee can become a "contractual teacher". After passing an examination, he/she can attain the status of "appointed teacher". The fourth level— "chartered teacher"—is the last level for most teachers. Some, however, manage to reach the fifth, honorary level, which is "professor of education". Surprisingly, the stage of professional promotion connected with the number of years of employment has no obvious influence on whether or not teachers integrate environmental protection issues into their curricula (Table 7). Of the five professional tiers, the highest proportion of teachers indicating their coverage of these topics in their classes is the group of chartered teachers (almost 80%). The differences between teachers in the other stages are minor.

Table 7. Integration of environmental protection into curricula and stage of teachers' professional promotion.

| Trainee | Contractual Teacher | Appointed Teacher | Chartered Teacher | Total |
|---|---|---|---|---|
| % | % | % | % | % |
| 69.23 | 64.44 | 65.85 | 79.08 | 73.59 |

Bigger differences are visible between teachers working in different locations—most often, environmental protection issues are addressed by educators in villages, while they are taught slightly less frequently in small towns and cities (Table 8).

Table 8. Location of the school and integration of environmental protection into curricula.

| Village | Small Town | Big City | Total |
|---|---|---|---|
| % | % | % | % |
| 80.47 | 69.12 | 70.59 | 73.59 |

## 3.2. Ways of Integrating Environmental Protection Issues into Curricula

Teachers who declared that they integrate topics related to environmental protection into their curriculum were asked to indicate how they introduce them in their work and how often they do it. Analysis of the results shows that the variable *subject taught* has the greatest influence on the way that environmental protection issues are integrated.

About 60% of the teachers declaring that they address the issue of environmental protection during their classes very often do soby discussing the topic during their subject lessons. What is surprising is that a very similar percentage of foreign language and science and natural sciences teachers declared that they use this method very often (68.92% and 69.77%, respectively—see Table 9). This means that it is not only the teachers of science and natural sciences who effectively incorporate issues and topics related to the environment into their daily work: foreign language teachers also willingly address these topics with their students.

Table 9. Covering ESD-related topics during subject lessons and subject taught.

| | Polish Language | Humanities and Social Sciences | Foreign Languages | Mathematics | Natural Sciences | Information and Technical Subjects | Total |
|---|---|---|---|---|---|---|---|
| Never | 3.33% | 4.48% | 4.05% | 11.11% | 0.00% | 0.00% | 3.56% |
| Occasionally | 1.67% | 2.99% | 1.35% | 13.89% | 3.49% | 0.00% | 4.15% |
| Sometimes | 35.00% | 29.85% | 25.68% | 30.56% | 26.74% | 44.44% | 29.08% |
| Very often | 60.00% | 62.69% | 68.92% | 44.44% | 69.77% | 55.56% | 63.20% |

Of all the teaching disciplines, the lowest proportion of teachers who declared that they address environmental protection issues during their lessons are the mathematicians. One in 10respondents never do it; almost 15% do it occasionally; one-third indicated "sometimes", and 44.44% chose "very often". Teachers of Polish and other human and social sciences include these topics among the problems discussed during their classes.

Another applied method of teaching environmental protection principles is by addressing them during additional classes, e.g., interest clubs. We need to point out that, in Poland, extra classes in schools are not popular. Teachers are not obliged by the law or school managers to engage in additional activities. Some of them, especially when they apply for promotion, decide to lead extra classes. It also depends on school policy—sometimes, directors go to great effort to offer as many extra activities as possible.

The above approach turns out to be a much less popular method of integrating environmental issues into school the curricula (Table 10). About one-fifth of Polish, foreign languages, social and human sciences teachers and one-fourth of science and natural sciences teachers impart ESD principles in this way. Only 1in 10mathematicians or ICT and technics teachers address these issues very often during additional classes. It is noteworthy that as many as one-fourth of surveyed mathematicians never discuss environmental protection issues during their extra classes. Similarly, almost 20% of Polish teachers declared that they never address these problems in their additional classes. About 10% of human and social sciences, foreign language, and natural sciences teachers never address these issues during the extra classes they lead.

Table 10. Covering education for sustainable development (ESD)-related topics during additional classes, e.g., interest clubs, and subjects taught.

|  | Polish Language | Humanities and Social Sciences | Foreign Languages | Mathematics | Natural Sciences | Information and Technical Subjects | Total |
|---|---|---|---|---|---|---|---|
| Never | 18.33% | 10.45% | 12.16% | 25.00% | 12.79% | 0.00% | 13.95% |
| Occasionally | 11.67% | 14.93% | 18.92% | 22.22% | 19.77% | 44.44% | 18.10% |
| Sometimes | 48.33% | 53.73% | 45.95% | 41.67% | 40.70% | 44.44% | 45.70% |
| Very often | 21.67% | 20.90% | 22.97% | 11.11% | 26.74% | 11.11% | 22.26% |

Teachers may implement the principles of environmental protection by assigning homework. However, it turns out that this is not a popular way of integrating these issues into curricula (Table 11). Almost 40% of mathematicians never do it. One-fifth of Polish language teachers declared that they never give home assignments connected with the issues of environmental protection (very often—only 5%). Teachers of subjects usually associated with this area—science and natural sciences—only sometimes (40.7%) or occasionally (34.88%) give home assignments regarding environmental protection. It is surprising because if teachers working with youths aged 13–20 want to raise students' awareness of sustainable development, they can expect that students would actually work independently at home.

Table 11. Assigning homework connected with the areas of sustainable development and subject taught.

|  | Polish Language | Humanities and Social Sciences | Foreign Languages | Mathematics | Natural Sciences | Information and Technical Subjects | Total |
|---|---|---|---|---|---|---|---|
| Never | 21.67% | 16.42% | 16.22% | 38.89% | 15.12% | 11.11% | 18.99% |
| Occasionally | 23.33% | 31.34% | 32.43% | 41.67% | 34.88% | 22.22% | 31.75% |
| Sometimes | 50.00% | 41.79% | 43.24% | 11.11% | 40.70% | 66.67% | 40.95% |
| Very often | 5.00% | 10.45% | 8.11% | 8.33% | 9.30% | 0.00% | 8.31% |

A quite popular method of introducing environmental protection issues is by encouraging students to engage in out-of-school activities, for example, joining different actions or events (Table 11). We need to emphasize that there are many local NGOs in Poland that promote the principles of sustainable development in different social groups. This provides teachers with great opportunities to promote these principles among their students. It is an attractive way that does not require much effort from teachers. Additionally, as proved by international research [33], the level of involvement and awareness of educational goals that are easily achieved by out-of-school activities translate into the effective implementation of these goals among students.

Only 1 in 10 Polish language, human and social sciences, and foreign languages teachers never use this approach to integrating environmental protection issues into their curricula (Table 12). More than 40% of Polish, science and natural sciences, and human and social sciences teachers declare that they very often encourage their students to engage in out-of-school activities connected with environmental protection. Over 16% of mathematicians never encourage students to engage in extra activities for the purpose of ESD, but slightly more than 30% very often do (44.44%—sometimes).

Table 12. Encouraging students to take up out-of-school activities/engage in different actions and subject taught.

|  | Polish Language | Humanities and Social Sciences | Foreign Languages | Mathematics | Natural Sciences | Information and Technical Subjects | Total |
|---|---|---|---|---|---|---|---|
| Never | 10.00% | 8.96% | 10.81% | 16.67% | 6.98% | 0.00% | 9.50% |
| Occasionally | 8.33% | 8.96% | 24.32% | 8.33% | 13.95% | 11.11% | 13.35% |
| Sometimes | 38.33% | 40.30% | 32.43% | 44.44% | 37.21% | 33.33% | 37.39% |
| Very often | 43.33% | 41.79% | 32.43% | 30.56% | 41.86% | 55.56% | 39.76% |

*3.3. Relations between Ways of Integrating Environmental Issues into Curricula and Selected Variables*

Using Pearson's chi-square test, we tried to determine whether there are any statistically significant relations between independent variables and the ways that teachers introduce environmental protection issues into curricula. Statistically significant results are those with $p < 0.05$.

Analysis of the results reveals that there is a statistically significant correlation between including the issue of environmental protection into subject curricula and school location (Tables 13–15).

Table 13. Integration of ESD-related topics during subject lessons and school location.

|  | Location | | | |
|---|---|---|---|---|
|  | Village | Small Town | Big City | Total |
|  | % | % | % | % |
| Never | 2.34% | 2.94% | 5.88% | 3.31% |
| Occasionally | 2.34% | 3.68% | 8.82% | 4.22% |
| Sometimes | 20.31% | 29.41% | 42.65% | 28.61% |
| Very often | 75.00% | 63.97% | 42.65% | 63.86% |
| Total | 100.00% | 100.00% | 100.00% | 100.00% |

Table 14. Chi-square tests.

|  | Value | df | Asymptotic (Bilateral) Significance |
|---|---|---|---|
| Pearson chi-square | 21.428 [a] | 6 | 0.002 |
| Likelihood ratio | 20.838 | 6 | 0.002 |
| Linear relationship test | 16.344 | 1 | 0.000 |
| N of significant observations | 332 | | |

[a] 33.3% of cells (4) have an expected frequency of less than 5. Minimum expected frequency is 2.25.

Teachers in village schools more often implement environmental protection issues in their subject lessons. This may mean that in rural areas, where students have regular contact with nature, teachers are more aware of the needs of our planet and are more willing to include such issues as using renewable sources of energy or recycling in their daily curricula. It is unexpected that teachers working in cities, where there is intensive anti-smog campaigning [33] and heavy use of renewable energy sources, less often introduce environmental protection issues into their daily lessons.

**Table 15.** Symmetric measurements.

|  |  | Value | Approximate Significance |
|---|---|---|---|
| Nominal by nominal | Phi | 0.254 | 0.002 |
|  | Cramer's V | 0.180 | 0.002 |
| N of significant observations |  | 332 |  |

Another statistically significant relationhipis the relation between including environmental protection issues into curricula and thedegree of professional promotion (Tables 16–18).

**Table 16.** Coverage of ESD-related topics during subject lessons and level of professional promotion.

|  | Degree of Professional Promotion | | | | |
|---|---|---|---|---|---|
|  | Trainee | Contractual Teacher | Appointed Teacher | Chartered Teacher | Total |
|  | % | % | % | % | % |
| Never | 23.08 | 4.44 | 2.44 | 2.55 | 3.57 |
| Occasionally | 0.00 | 8.89 | 4.88 | 3.06 | 4.17 |
| Sometimes | 7.69 | 31.11 | 29.27 | 29.59 | 28.87 |
| Very often | 69.23 | 55.56 | 63.41 | 64.80 | 63.39 |
| Total | 100.00 | 100.00 | 100.00 | 100.00 | 100.00 |

**Table 17.** Chi-square tests.

|  | Value | df | Asymptotic (Bilateral) Significance |
|---|---|---|---|
| Pearson chi-square | 21.140 [a] | 9 | 0.012 |
| Likelihood ratio | 14.363 | 9 | 0.110 |
| Linear relationship test | 3.875 | 1 | 0.049 |
| N of significant observations | 336 |  |  |

[a] 43.8% of cells (7) have an expected frequency of less than 5. Minimum expected frequency is 0.46.

**Table 18.** Symmetric measurements.

|  |  | Value | Approximate Significance |
|---|---|---|---|
| Nominal by nominal | Phi | 0.251 | 0.012 |
|  | Cramer's V | 0.145 | 0.012 |
| N of significant observations |  | 336 |  |

As shown by the research results, the youngest teachers implement environmental protection issues into curricula the most often. This means that newly graduated teachers are the ones who are the most aware of the need to implement environmental protection. This, in turn, may mean that universities are training teachers to focus on ESD exercises, including environmental protection, using renewable energy sources, recycling, etc.

There is a statistically significant correlation between the level of education taught and the incorporation of the issue of environmental protection into subject curricula during additional classes (Tables 19–21).

**Table 19.** Coverage of ESD-related topics during additional classes, for example, interest clubs, and type of school.

|  | School Type | | |
|---|---|---|---|
|  | Lower Secondary | Upper Secondary | Total |
|  | % | % | % |
| Never | 10.73 | 18.12 | 14.11 |
| Occasionally | 12.43 | 24.16 | 17.79 |
| Sometimes | 51.98 | 38.26 | 45.71 |
| Very often | 24.86 | 19.46 | 22.39 |
| Total | 100.00 | 100.00 | 100.00 |

**Table 20.** Chi-square tests.

|  | Value | df | Asymptotic (Bilateral) Significance |
|---|---|---|---|
| Pearson chi-square | 13.771 [a] | 3 | 0.003 |
| Likelihood ratio | 13.807 | 3 | 0.003 |
| Linear relationship test | 9.014 | 1 | 0.003 |
| N of significant observations | 326 | | |

[a] 0.0% of cells (0) have an expected frequency of less than 5. Minimum expected frequency is 21.02.

**Table 21.** Symmetric measurements.

|  |  | Value | Approximate Significance |
|---|---|---|---|
| Nominal by nominal | Phi | 0.206 | 0.003 |
|  | Cramer's V | 0.206 | 0.003 |
| N of significant observations |  | 326 |  |

The result of the analysis shows that the practical integration of environmental protection issues into additional classes depends on the level of education taught—it is more often introduced by teachers in lower-secondary schools. This may result from the fact that extra classes in upper-secondary schools are focused on preparing students to take their maturity exam (in Poland, high school graduates take an obligatory exam, the so-called *matura*, which is necessary to enter university). Thus, teachers in lower-secondary schools have more time and opportunities to address the problems of sustainable development and environmental protection during their additional lessons.

The last statistically significant relation is between the integration of environmental protection issues into curricula by encouraging students to take up out-of-school activities/engage in different actions and school location (Tables 22–24).

**Table 22.** Encouraging students to take up out-of-school activities/engage in different actions and subject taught, and school location.

|  | Location | | | |
|---|---|---|---|---|
|  | Village | Small Town | Big City | Total |
|  | % | % | % | % |
| Never | 4.69 | 11.76 | 13.24 | 9.34 |
| Occasionally | 7.81 | 15.44 | 19.12 | 13.25 |
| Sometimes | 39.06 | 34.56 | 39.71 | 37.35 |
| Very often | 48.44 | 38.24 | 27.94 | 40.06 |
| Total | 100.00 | 100.00 | 100.00 | 100.00 |

Table 23. Chi-square tests.

|  | Value | df | Asymptotic (Bilateral) Significance |
|---|---|---|---|
| Pearson chi-square | 15.377 [a] | 6 | 0.018 |
| Likelihood ratio | 16.274 | 6 | 0.012 |
| Linear relationship test | 13.284 | 1 | 0.000 |
| N of significant observations | 332 | | |

[a] 0.0% of cells (0) have an expected frequency of less than 5. Minimum expected frequency is 6.35.

Table 24. Symmetric measurements.

| | | Value | Approximate Significance |
|---|---|---|---|
| Nominal by nominal | Phi | 0.215 | 0.018 |
| | Cramer's V | 0.152 | 0.018 |
| N of significant observations | | 332 | |

It turns out that, like in case of extra classes, teachers in village schools most often impartenvironmental protection issues by encouraging students to engage in out-of-school activities. This means that teachers feel responsible for the protection of the natural environment and are more engaged in activities toward using renewable energy sources, recycling, or pollution prevention than their fellow professionals in towns or big cities.

## 4. Conclusions and Discussion

Schools, as institutions for general education, are believed to have a responsibility to equip their students with the knowledge and commitment to make personally meaningful decisions and act accordingly to address the challenges posed by both lifestyle and societal conditions. Achieving this goal requires, among other things, the adequate integration of these 'challenges' into the school curricula [34,35]. The research results reveal that Polish teachers working with youth aged 13–20 years in schools in Malopolska are aware that it is necessary to integrate the problems of environmental protection into a didactic process. They realize it is necessary to equip students with knowledge and skills regarding their surrounding environment and some global environmental aspects (general condition, threats), as it determines the future condition of our natural world. It is particularly interesting and important that it is not only teachers of subjects dedicated to the environment, like biology or physics, who are aware of the need to communicate these issues and who indicate thatthey address them in their curricula in many ways.

It is very important that teachers try to include environmental protection topics in their subject lessons. Foreign language teachers seem to be the most aware in this area. A possible reason is that because they know foreign languages, they more often read foreign (other than Polish) publications about ESD: for example, the ones created by UNESCO experts. It is also worth noting the fact that while working with adolescent students, these teachers can assign homework which shapes language competencies and are, at the same time, connected with environmental protection issues. It is extremely important to model active research attitudes among students, conviction about individual responsibility for the environment, and awareness of the threats resulting from human activity. Students should choose their actions based on sustainable development principles. Thus, teachers are responsible for ensuring that students are aware of how their actions affect the natural environment. It is possible that teachers will shape the knowledge and attitudes of students by integrating environmental protection issues into their curricula.

The analysis of the results shows that the more experienced the teachers, the more willing they are to promote sustainable development. Perhaps training and additional education of teachers who want to be promoted effectively raise their awareness and expand their knowledge about ESD. It is also possible that teachers who have worked longer are more aware of the necessity of integrating issues into their curricula, as suggested by UN education experts. For more experienced teachers, it is also easier to adopt different initiatives, like integrating key environmental issues into their didactic practice.

Teachers are less willing to address environmental protection issues during their additional classes, for example, interest clubs. Giving up this method of integrating key sustainable development problems into curricula may mean less effective education in the area of environmental protection.

We have presented the point of view of teachers regarding the integration of environmental protection issues into curricula, and the survey was conducted among educators in 1of the 16 administrative regions in Poland. It is worth looking at the research carried out in by Hidayah Liew Abdullah, Hamid, Shafii, Ta Wee, and Ahmad in Malaysia [1]. The results indicate that pupils' perception of the environment is overall high, and the location of the school does not significantly affect the pupils' perceptions of the environment. The majority of students define the concept of the environment as objects. In addition, the percentage of pupils who rationally regard the concept of the environment is also high. The perceptions of pupils on the need to protect the environment illustrate that the majority agree on the benefits it will impart to future generations. The environmental concepts are presented somewhat rationally and show the relationship between the environment and human beings. In conclusion, students' perceptions of the environment are positive; they can understand it rationally and know the nature of the natural disasters resulting from human actions.

It is also worth mentioning the research carried out in Mexico which indicated that official documents like general study plans, study programs of grades, and official students' textbooks include competences related to environmental education called 'competences for the coexistence', promoting harmonic relationships with others and with nature. Additionally, EE for sustainability was included as a transversal topic, with the expectation that participation will contribute to the graduate profile of basic education, indicating that students should promote and assume the care of health and the environment. Nevertheless, it is not clear how teachers are expected to implement EE in practice, meaning there is a lack of clarity inthe environmental theory that supports this curriculum [36].

Salmani, Hakimzadeh, Asgari, and Khaleghinez had conducted research in the field of environmental education in Iranian school curriculum [37]. The results show that the subjects presented at the highest frequency in sixth-grade social studies and science textbooks are ecology and human activities and the environment, respectively. Furthermore, human activities and the environment are more prominent in the science textbook rather than the social studies ones, although ecology is more prevalent in the latter. In terms of informational load and importance in these textbooks, ecology, the environment, and human activity are the main foundation of environmental education, respectively. In sum, in these textbooks, the three aforementioned bases have not been presented in parallel, and some modifications to the content of these textbooks are required to make them more understandable to the Iranian students.

The issue of integrating the key environmental protection aspects into curricula is also important in Serbia. In 2014, Stanišić and Maksić [38] published research results suggesting that students do not have enough knowledge to contribute to the development of a healthy lifestyle and environmental awareness. The latest changes in school policy and curricula confirm that the relevance of environmental education is recognized, but changes in school practice are yet to come. Also, Jukić [39], in his research, showed that ecological content currently dominates in the natural sciences among the school subjects. Additionally, both students and teachers graded the suitability of ecological content higher in relation to its current degree of representation inall school subjects. These results show that there is room and interest within the educational system (of both students and teachers) for the increase ine cological content in all school subjects. We should not expect young people to independently connect the pieces of information they receive from their teachers but should rather work on changing the educational system. The essential elements of ecology as a scientific discipline are so complex that, in order to reach a more comprehensive understanding, we cannot rely only on the natural sciences viewpoint. In other words, we need to teach ecology as a whole, as a system composed of proto-elements which appear in all school subjects throughout school curricula.

In their research, Molapo, Stears, and Dempster [40] indicated that, in Lesotho schools, the intended curriculum contains laudable goals with regard to learning in and about the environment. However, teachers interpret the curriculum in such a way that they teach mainly about the environment and never in the environment, and they seldom engage learners in activities in which they could develop positive attitudes that would encourage them to act with consideration of the environment.

As we can see, the adaptation of curricula to the aims of ESD is the object of many studies and scientific interest all over the world. Numerous research results prove that school curricula are crucial for promoting the principles of sustainable development.

## 5. Recommendations

Sustainable development aimsto ensure the improvement of the quality of life and economic progress without negatively affecting the environment o rreducing the availability of natural resources for future generations. According to UNESCO experts, sustainable development must be supported by properly planned and effectively implemented education. For ESD principles to be taught in practice and, first of all, known, the issues of sustainable development must be an obligatory part of teacher training programs. Teachers must realize that these principles are necessary and must know how to practically integrate sustainable development issues into their curricula.

In Poland, there are numerous discussions on introducing environmental *protection* to universities as an obligatory subject by means of a regulation issued by the Ministry of Science and Higher Education. It would be compulsory for economic, human sciences, and art studies, environmental engineering, and, first of all, for all pedagogy curricula (all teaching specializations). It seems to be a reasonable move, as teachers' actions would then be intentional (not incidental), and they would be more willing to integrate sustainable development content into their curricula (not only curricula of biology, chemistry, or natural sciences).

Results of the analysis show that foreign language teachers more often and more willingly address sustainable development issues during their subject classes. This means that other teachers may not have enough knowledge in this area, which, in turn, may mean they do not know of the publications addressing the problems of sustainable development. Therefore, teachers need to receive materials and tools in Polish, which would help them to understand the assumptions and principles of sustainable development and become familiar with the ways of incorporating ESD-related issues into their curricula.

Education for sustainable development in Poland is supported by EU funding. In 2016, the Ministry of Environment accepted rankings in two competitions of the Infrastructure and Environment Operational Programme for Measure 2.4 Environment protection and the promotion of ecological habits, sub-measure 2.4.5 Information and education activities in the area of environmental protection and effective use of natural resources. Forty-four applications were submitted to the competition POIS/2.4.5/2/2015 type b: Building the potential and integration, with a total value of 44 million PLN in EU support. Seventeen applications, for a total value of 18 million PLN, received s positive formal and quality assessment. The financial allocation of the competition was about 13 million PLN. The projects involve education through new media. They are examples of implementing good practices which significantly increase students' and teachers' awareness and knowledge about sustainable development issues.

In some Polish schools, compliance with sustainable development principles is seen as a school mission. Sustainable development topics are not only addressed in subject curricula but also in competitions, projects, the organization of the school premises, and the engagement in the protection of the local environment. It is a very good example of the practical implementation of sustainable development principles, as it not only raises awareness in students, teachers, and the local community but also enables the active promotion of sustainable development.

It is very important to connect the efforts made by schools to promote sustainable development with out-of-school activities, e.g., in cooperation with local communities. Such integrated actions will be more effective not only for students but also for the community [40].

Only joint activities of schools, institutions that educate and train teachers, the local community, and media are likely tolead to the full incorporationof environmental protection issues in school curricula. Therefore, we need to make every effort to enable such activities [41].

**Author Contributions:** Conceptualization, A.M. and I.O.; Data curation, A.M.; Formal analysis, I.O.; Methodology, A.M., I.O.; Writing – review & editing, K.W.-S.

**Funding:** This research received no external funding.

**Conflicts of Interest:** The authors declare no conflict of interest.

## References

1. Alrikabi, N.K.M.A. Renewable Energy Types. *J. Clean Energy Technol.* **2014**, *2*, 61–64. [CrossRef]
2. Boeve-de Pauw, J.; Gericke, N.; Olsson, D.; Berglund, T. The Effectiveness of Education for Sustainable Development. *Sustainability* **2015**, *7*, 15693–15717. [CrossRef]
3. Abdullah, N.H.L.A.; Hamid, H.; Shafii, H.; Ta Wee, S.; Ahmad, J. Pupils Perception Towards the Implementation of Environmental Education Across Curriculum in Malaysia Primary School. *J. Phys. Conf. Ser.* **2018**, *1049*, 012098. [CrossRef]
4. Pearce, D.; Barbier, E.; Markandya, A. *Sustainable Development.Economics and Environment in the Third World*; Earthscan: London, UK; Washington, DC, USA, 2010.
5. Kuzior, A. Dekada Edukacjidla Zrównoważonego Rozwoju. *Zeszyty Naukowe Politechniki Śląskiej* **2014**, *72*, 87–100.
6. Leicht, A.; Heiss, J.; Byun, W.J. *Issues and trends in Education for Sustainable Development*; UNESCO Publishing: Paris, France, 2018.
7. Frisk, E.; Larson, K.L. Educating for Sustainability: Competencies & Practices for Transformative Action. *J. Sustain. Educ.* **2011**, *2*, 1–20.
8. United Nations Educational, Scientific and Cultural Organization. *Education for Sustainable Development: Learning Objectives*; United Nations Educational, Scientific and Cultural Organization: Paris, France, 2017.
9. Jutvik, G.; Liepina, I. (Eds.) *Education for Change: A Handbook for Teaching and Learning Sustainable Development*; Gsndrs: Riga, Latvia, 2007.
10. Ocetkiewicz, I.; Tomaszewska, B.; Mróz, A. Renewable energy in education for sustainable development. The Polish experience. *Renew. Sustain. Energy Rev.* **2017**, *80*, 92–97. [CrossRef]
11. UNESCO. *Educational Module on Conservation and Management of Natural Resources*; Unesco-UNEP International Environmental Education Programme: Paris, France, 1986.
12. Tomaszewska, B. Sustainable Energy: Human Factors in Geothermal Water Resource Management. In *Advances in Human Factors in Energy: Oil, Gas, Nuclear and Electric Power Industries*; Springer: Cham, Switzerland, 2018.
13. Tomaszewska, B.; Rajca, M.; Kmiecik, E.; Bodzek, M.; Bujakowski, W.; Wątor, K.; Tyszer, M. The influence of selected factors on the effectiveness of pre-treatment of geothermal water during the nanofiltration process. *Desalination* **2017**, *406*, 74–82. [CrossRef]
14. Tomaszewska, B.; Pająk, L.; Hołojuch, G. Energy and environmental analysis of disposing of concentrate by injecting it back into the deep geological formation. *Desalin. Water Treat.* **2017**, *69*, 316–321. [CrossRef]
15. Robinson, K.; Aronica, L. *Creative Schools: The Grassroots Revolution That's Transforming Education*; Penguin Books: New York, NY, USA, 2016.
16. UNESCO. *Second Collection of Good Practices Education for Sustainable Development*; UNESCO Associated Schools: Paris, France, 2009.
17. UNESCO. *Global Action Programme on Education for Sustainable Development*; UNESCO Publishing: Paris, France, 2018; Available online: http://unesdoc.unesco.org/images/0024/002462/246270e.pdf (accessed on 16 May 2018).
18. Szempruch, J. *Nauczyciel w warunkachzmianyspołecznej*; IMPULS: Kraków, Poland, 2013.

19. Palka, S. *Nauczyciel jako badacz pedagogiczny [w:] J. Kuźma, J. Morbitzer (red.)*; Naukipedagogicznew teoriiipraktyceedukacyjnej: Kraków, Poland, 2003.
20. Schön, D.A. *The Reflective Practitioner. How Professionals Think in Action*; Basic Books: New York, NY, USA, 1987.
21. Giroux, H. Theories of Reproduction and Resistance in the New Sociology of Education. *Harv. Educ. Rev.* **1983**, *53*, 257–293. [CrossRef]
22. Liston, D.P. On Facts and Values. An Analysis of Radical Curriculum Studies. *Educ. Theory* **1986**, *36*. [CrossRef]
23. Parsons, T. The School class as a Social System. Some of its Functions in American Society. *Harv. Educ. Rev.* **1959**, *29*, 297–318.
24. McLaren, P. *Schooling as Ritual Performance: Towards a Political Economy of Educational Symbols and Gestures*; Rowman & Littlefield: Boston, MA, USA, 1986.
25. Bardziejewska, M. Okres dorastania. Jak rozpoznać potencjał nastolatków? In *Psychologiczne portrety człowieka; Brzezińska*; Brzezińska Anna, I., Ed.; GWP: Gdańsk, Poland, 2015.
26. Bee, H.; Boyd, D. *Lifespan Development*; Pearson: New York, NY, USA, 2011.
27. UNESCO. *Transforming Our World: The 2030 Agenda for Sustainable Development*; UNESCO Publishing: Paris, France, 2015.
28. Babbie, E. *The Practice of Social Research*; Thomson Wadsworth: Belmont, MA, USA, 2001.
29. Kivunja, C.; Kuyini, A.B. Understanding and Applying Research Paradigms in Educational Contexts. *Int. J. High. Educ.* **2017**, *6*, 26–41. [CrossRef]
30. Christensen, B.; Johnson, L. *Educational Research: Quantitative, Qualitative, and Mixed Approaches*, 4th ed.; SAGE Publications: Thousand Oaks, CA, USA, 2011.
31. Commission Regulation (EU) No 651/2014 of 17 June 2014 Declaring Certain Categories of Aid Compatible with the Internal Market in Application of Articles 107 and 108 of the Treaty (CELEX: 32014R0651). Available online: https://www.paih.gov.pl/files/?id_plik=11991 (accessed on 18 May 2018).
32. Hattie, J. *Visible Learning for Teachers: Maximizing Impact of Learning*; Routledge: London, UK, 2012.
33. Wilczyńska-Michalik, W.; Wilczyński, M. Air pollution in Cracow—A glance into the future from a historical perspective. *Geobalcanica* **2017**, 79–82. [CrossRef]
34. Dalelo, A. Loss of biodiversity and climate change as presented in biology curricula for Ethiopian schools: Implications for action-oriented environmental education [Etiyopyaokullarindayeralan biyolojimüfredati{dotless} konulari{dotless}ndaniklimdeğişikliğivebiyolojik-çeşitliliğinyokolmasi: Eylemyönelimliçevreeğitimineyönelikçi{dotless}kari{dotless}mlar]. *Int. J. Environ. Sci. Educ.* **2012**, *7*, 619–638.
35. Paredes-Chi, A.A.; Viga-de Alva, M.D. Environmental education (EE) policy and content of the contemporary (2009–2017) Mexican national curriculum for primary schools. *Environ. Educ. Res.* **2018**, *24*, 564–580. [CrossRef]
36. Salmani, B.; Hakimzadeh, R.; Asgari, M.; Khaleghinezhad, S.A. Environmental education in iranian school curriculum, A content analyses of social studies and science textbooks. *Int. J. Environ. Res.* **2015**, *9*, 151–156.
37. Stanišić, J.; Maksić, S. Environmental education in Serbian primary schools: Challenges and changes in curriculum, pedagogy, and teacher training. *J. Environ. Educ.* **2014**, *45*, 118–131. [CrossRef]
38. Jukić, R. Environmental education in grammar school curricula [Intercurriculäreransatzzurökologischenerziehung und bildung]. *Soc. Ekol.* **2014**, *22*, 221–245.
39. Molapo, L.; Stears, M.; Dempster, E. Interpretation and implementation of the environmental education curriculum: A case study of three lesotho schools [Interpretering en implementering van die omgewingsopvoeding-kurrikulum:'n gevallestudie van drieskole in lesotho]. *Acta Acad.* **2012**, *44*, 202–229.
40. Reed Jonson, J.A. *Education for Sustainable Development in Eco-Schools. Contextualised Stories from England and South Africa*; Lambert Academic Publishing: Saarbrucken, Germany, 2013.
41. Petty, G. *Teaching Today. A Practical Guide*, 3rd ed.; Nelson Thornes Limited: Chentelham, UK, 2004.

© 2018 by the authors. Licensee MDPI, Basel, Switzerland. This article is an open access article distributed under the terms and conditions of the Creative Commons Attribution (CC BY) license (http://creativecommons.org/licenses/by/4.0/).

Article

# Measuring Industrial Customer Satisfaction: The Case of the Natural Gas Market in Greece

Dimitrios Drosos [1], Michalis Skordoulis [2,*], Garyfallos Arabatzis [2], Nikos Tsotsolas [1] and Spyros Galatsidas [2]

[1] Department of Business Administration, School of Business, Economics and Social Sciences, University of West Attica, 12244 Egaleo, Greece; drososd@uniwa.gr (D.D.); ntsotsol@uniwa.gr (N.T.)
[2] Department of Forestry and Management of the Environment and Natural Resources, School of Agricultural and Forestry Sciences, Democritus University of Thrace, 68200 Orestiada, Greece; garamp@fmenr.duth.gr (G.A.); sgalatsi@fmenr.duth.gr (S.G.)
* Correspondence: mskordoulis@gmail.com

Received: 29 January 2019; Accepted: 25 March 2019; Published: 30 March 2019

**Abstract:** This aim of this paper is to measure industrial consumer satisfaction in the natural gas sector in Greece. By using the Multicriteria Satisfaction Analysis (MUSA) method, the paper measures industrial customer satisfaction based on criteria concerning the provided products and services, communication and collaboration with providers' staff, customer service, pricing policy and website. The research results that are based on the analysis of 95 questionnaires collected during the period between June 2017 and October 2017 show that the index of the global customer has a good performance as its value is about 74.99%. Furthermore, the satisfaction criterion with the highest performance is the one concerning communication and collaboration with natural gas providers' staff. It should be noted that the criterion concerning the provided products and services criterion is the only one with high performance and importance—meaning that it should be in the spotlight of the natural gas providers. The paper concludes that there is considerable space for improvements to be made. Customer satisfaction is of great importance for every company, as it can be highly connected with its performance. Using the results of this study, natural gas providers will have the chance to frame their future actions in order to keep their industrial customers satisfied. Taking into account both the fact that industrial customers' share in the Greek natural gas market is about 25% and that this market has been recently liberalized, it is of vital importance for natural gas providers to have sufficient information about their industrial customers' satisfaction.

**Keywords:** natural gas; energy market; customer satisfaction; industrial customers; multicriteria analysis

## 1. Introduction

It is widely known to both the companies and the consumers, that customer satisfaction, is one of the most significant factors that can guarantee success [1]. Day by day, consumers become increasingly more demanding for the quality of products and services they receive. They demand that the product or the service they pay for, fulfills their needs, is of high quality and is also offered at an affordable price.

Furthermore, it is a fact that every company, regardless of the field in which it operates, must find ways to keep its customers and attract new ones, while at the same time being competitive and profitable. As customers have access to more information nowadays, they become more flexible in their decisions and have more choices than ever before, it is even more important for companies to earn their trust. Furthermore, today's consumers have an educational level that allows them to judge any product or service they pay for, while they demand the best possible products and services at the same time. Consequently, a company that meets or even exceeds their customers' expectations has a significant competitive advantage [2].

Despite the fact that customer satisfaction is in the spotlight of many years of research, most literature focuses on the satisfaction concerning retail consumers instead of the industrial ones [3]. In recent years, most academic research on customer satisfaction has focused on consumer goods and services. Current studies concerning consumer goods and services typically relate satisfaction to a single discrete transaction [4–6]. On the other hand, studies concerning industrial customer satisfaction have emphasized the importance of customer–supplier relationships [7–9]. It is obvious that customer satisfaction in the industrial sector is understood as a relationship-specific rather than a transaction-specific construct [3,10].

Energy is one of the key regulators of the interaction between nature and humans. Many environmental issues that lead to negative consequences for society, the economy and the sustainability of the environment are connected with the production, transformation, and use of energy [11]. The efficient treatment of these issues is a vital necessity [12]. Therefore, the need for ecologically friendly and cleaner forms of energy is becoming more and more imperative [13,14]. As a result of the big changes in the global energy system coming from the above needs—which will take place in the following decades, renewable sources of energy are expected to be the grand victors in the race for the increasing energy demand supply [15–17]. Despite the fact that natural gas is not a renewable source of energy, it is relatively cleaner than its alternatives [18].

Thus, during the last decades, the demand for both renewable and cleaner sources of energy has increased considerably both on the retail and industrial customer level. The quality of the services provided by the clean energy supply companies and the satisfaction of their customers, have been particularly important for them during the last few years. Therefore, for the energy supply companies, the measurement of their customers' satisfaction is a contemporary tool for strategic planning, capable of creating the necessary conditions for their survival and development.

The main objective of the present research is to measure industrial customer satisfaction of natural gas providers in Greece using a multicriteria analysis method. The methodological approach is based on the principles of multicriteria modelling as the Multicriteria Satisfaction Analysis (MUSA) Method is used for the satisfaction measurement.

## 2. Literature Review

### 2.1. Industrial Customers' Satisfaction

Several studies show that the long-term success of an organization is closely related to its ability to adapt to its customer needs and changing preferences [19–22]. The analysis of customer satisfaction and its comparison with the results of similar studies can provide policy makers with a unique insight into the motivations and satisfaction of consumers [23–25].

Furthermore, customers have access to an educational level that allows them to judge any product or service they pay for, while they demand the best possible products and services at the lowest price. Consequently, a company that meets or even exceeds a customer's expectations has a significant competitive advantage [26–28]. Therefore, it is important for every company to know who its customers are, what they expect and how adequately their expectations are met. This process is of the same importance for both the retail and industrial consumers [29].

The theory of customer satisfaction has been presented in international literature, as a reliable tool for the evaluation of a company's results. Customer satisfaction under the view of perceived quality may be seen as the difference between the actual and the expected quality of a product or a service. Many researchers emphasize on satisfaction with functional attributes [30].

According to Ostrom and Iacobucci [31], customer satisfaction is a way to evaluate the gap between the expectations of a customer from a particular product or service, and what the customer receives after the use of said product. For the measurement of customer satisfaction certain indexes, price, efficiency or total performance of a company are being used. Woodside et al. [32], considers customer satisfaction the main factor that affects their behavior. Yi [33], defines customer satisfaction in two basic ways: either as a result or as a procedure. Furthermore, Yi [33] considers that the definition of customer satisfaction varies, depending on a series of definitive factors that concern the satisfaction from a product or service, a buying decision experience, a performance attribute, the end-user experience, a company shop or department and, the satisfaction from a pre-buying experience.

Churchill and Suprenant [6] state that *"satisfaction is the result of purchase and use of a product or service, which derives from the customer's comparison between the remuneration and the cost of purchase, taking into consideration the expected result"*. Customer satisfaction is also based on customer knowledge, which is about products, suppliers and markets [34].

Kotler [35], approaching the issue of customer satisfaction from a marketing point of view, has analyzed the meaning of customer satisfaction as feelings of pleasure or discontent, which can be derived from the comparison between the performance or the result of a product or service, and the expectations the person developed before paying for it.

In recent years, it has been widely recognized that customer satisfaction is an essential tool of corporate marketing strategies. According to O'Sullivan and McCallig [36], customer satisfaction has a positive impact on a firm's value. The researchers found that this impact is higher than the impact its earnings have on its value. They also found that customer satisfaction positively and significantly moderates the earnings–firm value relationship. Other researchers found that customer satisfaction has a direct and positive effect on customer purchase intentions and loyalty [37–39].

Based on the abovementioned analysis, we can conclude that high levels of customer satisfaction can lead to customer loyalty [40–42], business profitability [43,44], trust [45], customer retention [41], positive word-of-mouth [46,47], repeating sales [42,48], future revenues [49–54] increased stock prices [49,55–57], and higher market share [58,59].

During recent decades, most researchers on customer satisfaction have focused on satisfaction with consumer goods and services. As far as the satisfaction of industrial customers is concerned, research is still not particularly advanced [60].

Raj et al. [61] defines industrial customer satisfaction as a relationship-specific construct, describing how well a supplier meets a customer's expectation in the following areas: product-related information, services, complaint handling, order handling, product features, interaction with internal staff and interaction with salespeople.

Industrial customers' satisfaction is closely connected with the quality of the provided products and services provided to them, and is necessary for the continuous improvement and excellence of any company [62–64]. As is the case of retail customers, industrial customers' satisfaction is also connected with the performance of companies and the development of a competitive advantage [6]. However, industrial customer satisfaction is found to be much more complex [3].

Homburg and Rudolph [3] developed a valid customer satisfaction measure for industrial customers, the INDSAT model. The model consists of seven distinct satisfaction dimensions: products, salespeople, product-related information, order handling, technical services, interaction with internal staff, and complaint handling.

*2.2. Customer Satisfaction Surveys in Energy Sector*

Throughout the last years, several studies have been carried out concerning customer satisfaction with energy providers. Generally, satisfaction in the energy sector covers the quality of a number of services, such as the provision of a new connection, the billing, the handling of customer requests, and complaints. Customer satisfaction can be a significant motivating factor for energy providers. The main objective of an energy organization is to acquire quality products and services that satisfy its customers with measurable improvements to mission capability and operational support in at a fair and reasonable price [65].

Customers who are satisfied with the service quality they receive, tend to trust their energy provider. Customer satisfaction is a prominent theme in the energy sector. Satisfied customers have the potential to become loyal customers and to attract new customers to an energy provider [66,67].

Mutua et al. [68], formulated a general framework for the study of customer satisfaction in the energy sector (electricity, petroleum, biomass, and renewable energy) in Kenya, using the European Consumer Satisfaction Index (ECSI). They found that customer satisfaction is highest in the renewable energy sector at 74.71%, followed by the petroleum sector at 62.32% and the biomass sector at 61.82%. The electricity sector has the least satisfied customers at 53.06%.

J.D. Power and Associates [69] conducted a study to measure business customer satisfaction with gas utility companies in four U.S.A geographic regions (East, Midwest, South, and West). This research is based on responses from 10,635 U.S. companies that spend about $150 per month on natural gas. The study examines six satisfaction criteria; corporate citizenship, billing and payment, communications, price, customer service and field service. The overall satisfaction among business customers of gas utilities averages at 674 on a 1000-point scale.

Moreover, J.D. Power and Associates [70] using the same set of criteria, measures residential customers' satisfaction with their gas utility in the same geographic regions in the U.S.A. Retail gas customers were reportedly more satisfied (overall satisfaction 706 index points on a 1000-point scale), than the industrial ones.

Ipsos, London Economics, and Deloitte [71] conducted consumer market research on the functioning of retail energy markets in the European Union (EU). The research covers all European Union Member States, Norway, and Iceland. Across the European Union countries, 40% of survey respondents "strongly agreed" (scores 8 to 10) that their energy provider offered an overall high quality of service, 40% "agreed" (scores 5 to 7) and finally 15% "disagreed" (scores 0 to 4). In Greece, 30% of survey respondents "strongly agreed", 41% "agreed" while 26% "disagreed".

The American Customer Satisfaction Index (ACSI) analyzed customer satisfaction with investor-owned energy utilities serving U.S. residential customers (electric and natural gas service). According the ACSI Customer Satisfaction Reports [71] the natural gas remains the superlative energy source with an improved score of 78% (100-point scale). The Residential household satisfaction with electricity is lower at 75%.

Liu et al. [72], developed an evaluation index system of electric power customers' satisfaction based on the service blueprint theory. The service blueprint model is divided into 4 parts, which are customer behavior, front office staff behavior, back office staff behavior, and support process.

Chodzaza and Gombachika [73], focused on functional quality offered by the public electricity provider to its industrial customers within Southern Region of Malawi. The findings suggest that the service quality is poor, irrespective of demographic characteristics of the industrial customers. The industrial customers were dissatisfied with the availability of power and customer care services.

Jannadi and Al-Saggaf [74], conducted a study to measure the customer satisfaction of a typical energy provider in Saudi Arabia. The study revealed that the provider had high satisfaction scores in tangibles dimensions, but low ones in the dimensions of responsiveness and reliability.

Medjoudj et al. [75], analyzed the customer satisfaction of power users in Bejaia City, Algeria, using the Analytic Hierarchy Process (AHP) method. The obtained results indicate the advantage of investment to improve customer satisfaction and enterprise profitability.

Medjoudj et al. [75] used three multi-criteria decision making methods (cost benefit analysis, economic criteria inspired from game theory, and the analytic hierarchy process) to find that customer satisfaction with energy providers is expressed by the requirements of high quality of service at the lowest possible cost of electricity.

Ibáñez et al. [76] proposed a framework where retail customer satisfaction with energy providers is correlated with the dimensions of technical quality of the core services (supply interruption, and service re-establishment), technical quality of the peripheral services (information, consultation, and flexible contracts) and, service process quality (prompt service, politeness, and customer requests). Their results confirmed the direct impact of the examined dimensions on customer satisfaction. Following the same path, Hartman and Ibáñez [77], proposed a conceptual framework for the impact of satisfaction and switching costs on customer loyalty in energy markets. The main factors related to customer satisfaction according to their analysis are technical quality of the core services, technical quality of the peripheral services, service process quality, value added services, innovations, environmental and social commitment, and pricing policy [77]. Their results show that in order to increase customer satisfaction with energy providers and thus, indirectly, customer loyalty—service quality should be increased [77].

Price seems to be important for the industrial customers of energy providers. However, even in this case, it is shown that industrial customers' satisfaction has an important effect on price tolerance even when switching barriers exist [78]. This is supported by the finding that even in the case of high switching barriers, they are not big enough to retain dissatisfied industrial customers [78].

### 2.3. Natural Gas Market in Greece

The use of natural gas is a strategic choice for the European Union. By using natural gas, the goal is the transition to an economy based on more environmentally friendly sources of energy than oil, while also being the turning point to renewable sources of energy. The above mentioned data can support the fact that natural gas is the second most used type of energy in the European Union, while it takes the third position in world level (Figure 1).

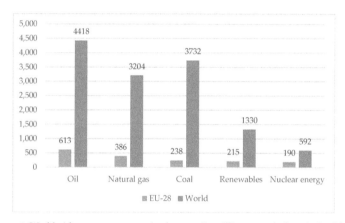

**Figure 1.** Worldwide energy consumption by type (in million tons of oil equivalent) [79].

As already mentioned, after oil (42%), natural gas is the most used type of energy in the European Union (23%). However, Greece is mostly based on oil consumption (66%), while natural gas consumption is equal to 4% of the total energy consumption balance (Figure 2). This shows that natural gas has a low level of penetration in the Greek market, meaning that there is enough space for growth.

To clearly describe the energy market of Greece, it is notable that the sectors of electricity and natural gas have undergone radical reforms during the last decade towards the full liberalization on the basis of European energy integration. The aim is both the sustainability of the energy system and the security of supply. Another important aspect of these reforms is the creation of conditions for the competitive functioning of the market. The right to choose an electricity supplier for all consumers has been established since 2008. Full liberalization of the natural gas market has also taken place since the beginning of 2018.

Greece fully depends on natural gas imports, with the greater proportion of its supplies coming from Russia, Algeria and Turkey. More specifically, Russian natural gas is 75% of total imports (DEPA). The Public Company of Gas Supply of Greece (DEPA) was found in 1988 and has been the main importer of natural gas pipes and liquefied natural gas (LNG) in Greece ever since. Total natural gas consumption in Greece, seems to follow the established trend of the European Union countries.

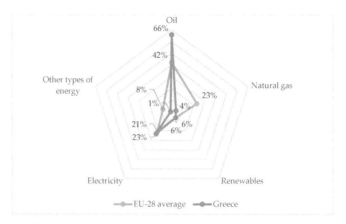

**Figure 2.** EU-28 average and Greece energy consumption by type (in % percent) [80].

Furthermore, according to Eurostat, natural gas prices in Greece follow the average prices of the European Union countries (including taxes and levies) as a decline has been recorded in recent years. More specifically, natural gas prices have decreased from €19.44 in 2012 to €16.41 in 2018 per gigajoule in the European Union countries, while in Greece they have decreased from €28.25 in 2012 to €14.79 in 2018 per gigajoule [80] (Figure 3).

Based on data provided by DEPA [81], the largest part of natural gas in Greece is used for electricity production (55%); industrial customers occupy a big percentage of the total market as their share is about 25%, while 19% of natural gas consumption is for home usage and 1% for vehicles.

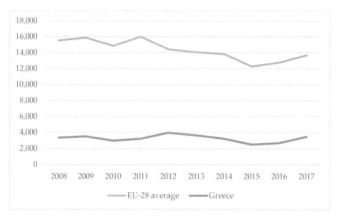

**Figure 3.** EU-28 average and Greece gross inland consumption of natural gas (in thousand tons of oil equivalent) [80].

## 3. Research Methodology

*3.1. Data and Research Tool*

The current study is based on a survey that took place between June and October 2017. The main research tool was a questionnaire, which was electronically distributed to 630 companies in order to be answered by the responsible managers for energy planning. We note that the managers who were asked occupied common positions, since according to literature [3,60] different roles may have a different model of preferences. Finally, 95 questionnaires were responded resulting in an overall response rate of approximately 15%. The survey focused on companies belonging to business sectors that account for about 60% of the total production of the Greek economy and use natural gas (Figure 4). The sample was randomly selected based on the data provided by published sectoral studies.

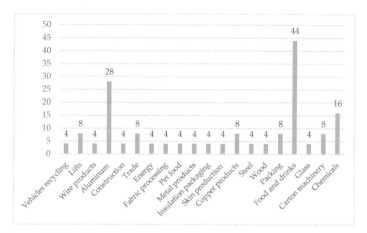

**Figure 4.** Number of companies took part in the survey categorized by industrial sector.

In order to design the questionnaire, the existing literature was examined. Based on the literature review and the specific characteristic of the natural gas market in Greece, a series of satisfaction criteria and sub-criteria emerged, as presented in Table 1.

**Table 1.** Satisfaction criteria and sub-criteria.

| Satisfaction Criteria | | Satisfaction Sub-Criteria | | |
|---|---|---|---|---|
| Products-services [3,61,65,68–70,75,77] | 1.<br>2.<br>3.<br>4. | Range of services<br>Connection costs<br>Connection process<br>Technical problems solving | 5. | Emergency response |
| Communication and collaboration with staff [3,60,61,65,68,72,76] | 1.<br>2.<br>3.<br>4. | Kindness & willingness<br>Knowledge & skills<br>Professionalism<br>Reliability | 5. | Confidentiality |
| Customer service [3,61,69,70,72–74,76] | 1.<br>2.<br>3.<br>4. | Promises keeping<br>Responses direct<br>Effectiveness<br>Information provision | 5.<br>6. | Absence of errors<br>Accounts accuracy |
| Pricing policy [3,60,65,69,70,75,77,78] | 1.<br>2.<br>3.<br>4.<br>5. | Charges<br>Value for money<br>Payment methods<br>Prices flexibility<br>Discounts | 6.<br>7. | Update changes<br>Payment terms |
| Website [61,72,76,77] | 1.<br>2.<br>3.<br>4. | Information<br>Navigation ease<br>Loading speed<br>Aesthetic design | 5. | Content |

A five point Likert scale was used to measure respondents' level of satisfaction rating from totally dissatisfied (1) to totally satisfied (5). Including the above mentioned satisfaction oriented questions, a total of 35 questions were answered by each of the respondents to the research.

The research data were analyzed using the Multicriteria Satisfaction Analysis (MUSA) method. The MUSA method which will be applied in the present research, uses satisfaction data collected through special questionnaires. Each respondent is asked through a questionnaire, to express satisfaction which depends on a set of criteria and sub-criteria [82].

### 3.2. An Overview of the Mulitcriteria Satisfaction Analysis (MUSA) Method

The MUSA (Multicriteria Satisfaction Analysis) method is a multicriteria preference disaggregation analysis technique [63,83]. The method assumes that customer's global satisfaction depends on criteria representing satisfaction dimensions; its main objective is the aggregation of judgments into a collective value function of satisfaction. The method is based on ordinal regression in order to facilitate that global satisfaction becomes as consistent as possible with customers' judgments.

More specifically, the method infers an additive collective value function Y* and a set of partial satisfaction functions $X_i^*$, given customer's global satisfaction Y and partial satisfaction $X_i$. The main objective of is to achieve the maximum consistency between the value function Y*and the judgments of Y. The method's ordinal regression function is following one:

$$\begin{cases} Y^* = \sum_{i=1}^{n} b_i X_i^* \\ \sum_{i=1}^{n} b_i = 1 \end{cases} \quad (1)$$

where Y* is the estimation of global satisfaction, n is the number of satisfaction criteria and $b_i$ is a positive weight of the i-th criterion.

In order to face the problem of the model's stability, an optimality analysis stage is included. The final solution is obtained by exploring the polyhedron of near optimal solutions and is calculated by a number of linear programs equal to the number of the satisfaction criteria:

$$\begin{cases} [\max] F\prime = \sum_{k=1}^{a_i - 1} w_{ik} \; \gamma\iota\alpha \; i = 1, 2, \ldots, n \\ \text{under the constraints:} \\ F \leq F^* + \varepsilon \end{cases} \quad (2)$$

where $\varepsilon$ is a small percentage of F*. The average of the solutions given by the n LPs may be taken as the final solution. In case of non-stability this average solution is less representative.

The assessment of a performance norm may be very useful in customer satisfaction analysis. The average global and partial satisfaction indices are used for this purpose and can be assessed according to the following equations:

$$S = \sum_{m=1}^{a} p^m y^{*m} \text{ and } S_i = \sum_{k=1}^{a_i} p_i^k x_i^{*k} \quad (3)$$

where S and $S_i$ are the average global and partial satisfaction indices, and $p^m$ and $p_i^k$ are the frequencies of customers belonging to the $y^m$ and $x_i^{*k}$ satisfaction level, respectively.

### 4. Research Results

Table 2 summarizes the reliability of the scales. The results show that all the dimensions used in the research were highly reliable based on Cronbach's a values.

Table 2. Satisfaction criteria reliability.

| Satisfaction Criteria | Number of Sub-Criteria | Cronbach's a |
|---|---|---|
| Products–services | 5 | 0.801 |
| Communication and collaboration with staff | 5 | 0.710 |
| Customer service | 6 | 0.725 |
| Pricing policy | 7 | 0.823 |
| Website | 5 | 0.703 |

As stated above, satisfaction measurement is based on five main criteria. Approaching the importance of these criteria, shaped by the answers of industrial customers, it is shown that the criterion with the greatest importance is products and services (24.46%), while website (17.9%) is the one with the least importance (Table 3). These results are in accordance with the results of previous studies where products and services seem to be very important for industrial customers [76,77]. Furthermore, we may see that pricing policy seems to be of low importance (18.10%) (Table 3). This can be the result of the fact that the price of natural gas in Greece is lower than the EU-28 average.

Table 3. Satisfaction criteria weights.

| Satisfaction Criteria | Level of Importance for Global Satisfaction |
|---|---|
| Products–services | 24.50% |
| Communication and collaboration with staff | 19.60% |
| Customer service | 19.90% |
| Pricing policy | 18.10% |
| Website | 17.90% |

The criterion with the highest level of satisfaction is this of communication and cooperation with staff with an 86.46% percentage followed by customer service and products-services criteria with 85.31% and 78.99% levels of satisfaction respectively (Table 4).

Table 4. Satisfaction criteria indexes.

| Satisfaction Criteria | Level of Satisfaction |
|---|---|
| Products–services | 78.99% |
| Communication and collaboration with staff | 86.46% |
| Customer service | 83.51% |
| Pricing policy | 55.58% |
| Website | 70.47% |

Regarding industrial customers' level of global satisfaction, 6.82% of them were totally dissatisfied or dissatisfied, 27.27% were neutral, 40.91% were satisfied and 25% were totally satisfied (Table 5). In relation to retail customers' satisfaction as presented in the results of Ipsos, London Economics and Deloitte [84] analysis, industrial customers' satisfaction is lower. This is in line with the results of other studies where industrial customers have lower level of satisfaction than the retail ones [73].

Table 5. Industrial customers' global satisfaction.

| Satisfaction Scale | Level of Satisfaction |
|---|---|
| Totally dissatisfied | 4.55% |
| Dissatisfied | 2.27% |
| Neutral | 27.27% |
| Satisfied | 40.91% |
| Totally satisfied | 25% |

Based on Figure 5, the results of the survey are positive, as the average global satisfaction index of the customers is about 74.99%. According to the theory established by the MUSA method, the growth trend of the satisfaction function means that industrial customers are demanding about the products and services they receive [63,83]. These obtained results confirm the fact that industrial customers are more demanding than retail ones. Industrial customer demandingness is a critical factor for their suppliers as such customers can often get easily dissatisfied with the existing products and services so as they are searching for new to replace the old ones [85].

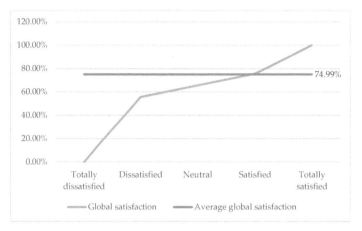

**Figure 5.** Global satisfaction function and average global satisfaction index.

Lastly, Figure 6 shows the strong and weak points of the natural gas providers examined. The criteria concerning pricing policy and website are located in the so-called "status quo area" of the action diagram. It should be noted that if these criteria show higher levels of significance in the future without improving the level of satisfaction, global satisfaction will get significantly worse.

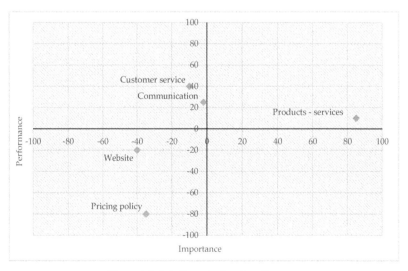

**Figure 6.** Satisfaction criteria action diagram.

The "products-services" criterion is located in the leverage opportunity area of the action diagram, which means that this criterion is of high performance and importance; thus, it is the criterion where the greatest attention should be paid. As far as it is concerned, this criterion should be in the spotlight of conservation actions and investments, as it could be characterized as a competitive edge for the market's companies.

Finally, the criteria "communication and collaboration with staff" and "customer service" are located in the transfer resources area which means that no funds should be invested for improving them, as they are of low importance.

The above mentioned results are useful in order to be used as a guidance for the examined companies' managers. However it is important to see if there is the need for a different approach to various industrial customers. In order to ascertain this assumption, a Kruskal Wallis test will be performed for using the industrial sector of the companies responded to the research as the independent variable (Table 6).

Table 6. Kruskal Wallis test's results.

| Satisfaction Criteria | Chi-Square | df | p-Value |
| --- | --- | --- | --- |
| Products–services | 9.412 | 4 | 0.082 |
| Communication and collaboration with staff | 3.877 | 4 | 0.423 |
| Customer service | 2.597 | 4 | 0.627 |
| Pricing policy | 3.170 | 4 | 0.530 |
| Website | 9.017 | 4 | 0.061 |
| Global satisfaction | 4.751 | 4 | 0.314 |

Based on Table 6 data, we see that in a 5% level of significance, there is no statistically significant difference between the levels of satisfaction and the industrial sector of the respondents. Thus, we may conclude that the above mentioned satisfaction criteria would be useful for measuring industrial customers' satisfaction belonging to the same sectors of the respondents to the research. Moreover, the strategy which must be followed by the natural gas providers based on the research results, could be the same for all the industrial customers of these sectors.

## 5. Discussion and Conclusions

The aim of this research was to measure industrial customer satisfaction with natural gas providers in Greece. Customer satisfaction can be a core determinant of the performance of any organization [1,3,86]. In the case of industrial customers, satisfaction can be of even greater importance due to the fact that such customers can easily seek new suppliers [85].

Research results show an adequate satisfaction level in the global satisfaction level about 74.99%. This level of satisfaction can be characterized as adequate, due to the fact that industrial customers are more demanding than retail ones and at the same time may have lower levels of satisfaction than the retail ones [73,85]. However, keeping in mind the fact that the percentage of neutral customers (either satisfied or dissatisfied) is around 27% of the total sample, a considerable space for improvement is recorded.

The need for improvement actions can be more obvious as in certain criteria such as this of pricing policy the level of satisfaction is very low. Nonetheless, the criteria with low performance are in the so called "status-quo" and are of the action diagram, which means they are not significantly important for industrial customers' satisfaction; thus, the natural gas providers should not invest very high resources in improving these criteria. Moreover, according to the MUSA method's results, the criterion located in the leverage opportunity area of the action diagram is this of products and services; thus, this is the most important criterion for industrial customer satisfaction and should be in the spotlight of natural gas providers.

These results could be used by energy providers' managers as a guidance for their strategy development. Measuring customer satisfaction as a useful insight on the performance of a company dealing with the needs of its customers can arise. The critical analysis of such results could significantly contribute to the disclosure of the actions that should be undertaken in order to increase customer satisfaction [3]. Such results are crucial for decisions concerning improvement actions and resources allocation as well.

We should note that the present research has some limitations. First, what should be taken into consideration for the results of the survey is that it has been carried out in the beginning of the natural gas market in Greece full liberalization period. This means that until the beginning of the survey, the choices in the market for alternative providers were relatively limited. This may also explain the fact that while satisfaction from criteria such as this concerning pricing policy is low, the level of demanding is low as well, in contrast to the results of other surveys [78]. This trend is most likely to change with the operation of a fully liberalized market. Especially in the case of natural gas, which is a strategic choice for the European Union and consequently for Greece, market growth margins and competition intensities will increase. Thus, the strategic importance of customer satisfaction will keep growing constantly [87].

Despite the fact that we see that the results do not differ between the different industrial sectors of the respondents, they should be tentatively accepted. This means that they cannot be automatically generalized [3] before a systematic replication is achieved [88]. Besides, the results of similar surveys show that there are differences in the satisfaction level of industrial customers belonging to different sectors and, as far as it is concerned, there is a need for different strategic choices [3,73].

Furthermore, this research is based on data which are not longitudinal offering a static view. In such cases as the examined one, it is important to use data concerning customer satisfaction over time in order to track over longer periods of time.

Last, the questionnaire used mainly included questions related to the measurement of industrial customers' satisfaction as the main objective was to calculate satisfaction indices. Factors related to customer satisfaction such the change rate of service providers, the purchase of added services, the willingness to pay or customers' loyalty are not examined. These factors are found to be strongly correlated with customer satisfaction in similar studies [15,73].

Based on the above analysis and the limitations noticed, there are several avenues for future research. First, more analytical research including the above mentioned factors related to customer satisfaction should be run. At the same time, the applied questionnaire could be updated with more criteria concerning industrial customer satisfaction and address different roles (e.g., deciders, buyers, and users) within the industrial customers.

Furthermore, a survey combining both individual consumers' and industrial customers' satisfaction would more effectively underline the strong points and the areas in need of improvement of the natural gas providers.

Finally, the installation of a permanent satisfaction barometer, which could be run annually, is an important extension as it could give the ability to track changes concerning both individual consumers' and industrial customers' satisfaction.

**Author Contributions:** D.D., M.S. and N.T. gathered the data and carried out the satisfaction measurement using the MUSA method. G.A. and S.G. gathered and implemented all the theoretical background of the paper. D.D., M.S. and N.T. carried out the connection between literature and the customer satisfaction analysis, as well as the finalization of the article.

**Funding:** This research received no external funding.

**Conflicts of Interest:** The authors declare no conflict of interest.

## References

1. Ahmad, N. Quality Attribute and Customer Satisfaction: Using Kano's Model to Prioritize What Matters Most to Customers. *J. Mark. Consum. Behav. Emerg. Mark.* **2017**, *1*, 15–28.
2. Skordoulis, M.; Alasonas, P.; Pekka-Economou, V. E-government services quality and citizens' satisfaction: A multi-criteria satisfaction analysis of TAXISnet information system in Greece. *Int. J. Prod. Qual. Manag.* **2017**, *22*, 82–100. [CrossRef]

3. Homburg, C.; Rudolph, B. Customer satisfaction in industrial markets: Dimensional and multiple role issues. *J. Bus. Res.* **2001**, *52*, 15–33. [CrossRef]
4. Pham, C.H. Customer satisfaction on service quality of consumer goods retailers: Evidence from Vietnam. *Int. J. Civ. Eng. Technol.* **2019**, *10*, 1159–1175.
5. Shin, D.H. Effect of the customer experience on satisfaction with smartphones: Assessing smart satisfaction index with partial least squares. *Telecommun. Policy* **2015**, *39*, 627–641. [CrossRef]
6. Churchill, G.A., Jr.; Surprenant, C. An investigation into the determinants of customer satisfaction. *J. Mark. Res.* **1982**, *19*, 491–504. [CrossRef]
7. Yashin, N.S.; Popova, L.F.; Bocharova, S.V.; Bagautdinova, N.G. Customer satisfaction assessment in management quality system of industrial enterprises. *Int. Bus. Manag.* **2016**, *10*, 5720–5726.
8. Festge, F.; Schwaiger, M. The Drivers of Customer Satisfaction with Industrial Goods: An International Study. *Adv. Int. Mark.* **2007**, *18*, 179–207.
9. Ismail, H.; Bakar, Z.; Salleh, A.H.M. Buyer satisfaction and loyalty - Evidence from the industrial goods market. *J. Pengur.* **2006**, *25*, 47–61.
10. Tong, L.; Hou, X.; Li, X. Empirical study on customer satisfaction influencing factors of industry application products based on experiential level theory. *China Commun.* **2016**, *13*, 260–268. [CrossRef]
11. Dincer, I. Environmental and sustainability aspects of hydrogen and fuel cell systems. *Int. J. Energy Res.* **2007**, *31*, 29–55. [CrossRef]
12. Papageorgiou, A.; Skordoulis, M.; Trichias, C.; Georgakellos, D.; Koniordos, M. Emissions trading scheme: Evidence from the European Union countries. In *Communications in Computer and Information Science*; Kravets, A., Shcherbakov, M., Kultsova, M., Shabalina, O., Eds.; Springer International Publishing: Cham, Switzerland, 2015; pp. 222–233.
13. Chalikias, M.S.; Kolovos, K.G. Citizens' views in Southern Greece part II. Contribution of forests to quality of life. *J. Environ. Prot. Ecol.* **2013**, *2*, 629–637.
14. Chalikias, M. Effect of natural recourses and socioeconomic features of tourists on the Greek tourism. *J. Environ. Prot. Ecol.* **2012**, *13*, 1215–1226.
15. Ntanos, S.; Kyriakopoulos, G.; Chalikias, M.; Arabatzis, G.; Skordoulis, M. Public perceptions and willingness to pay for renewable energy: A case study from Greece. *Sustainability* **2018**, *10*, 687. [CrossRef]
16. Ntanos, S.; Kyriakopoulos, G.; Skordoulis, M.; Chalikias, M.; Arabatzis, G. An Application of the New Environmental Paradigm (NEP) Scale in a Greek Context. *Energies* **2019**, *12*, 239. [CrossRef]
17. Skordoulis, M.; Chalikias, M.; Galatsidas, S.; Arabatzis, G. Competitive Advantage Establishment through Sustainable Environmental Management and Green Entrepreneurship: A Proposed Differential Equations Framework. In *Springer Earth System Sciences*; Theodoridis, A., Ragkos, A., Salampasis, M., Eds.; Springer International Publishing: Cham, Switzerland, 2019; pp. 205–219.
18. Di Pascoli, S.; Femia, A.; Luzzati, T. Natural gas, cars and the environment. A (relatively) 'clean' and cheap fuel looking for users. *Ecol. Econ.* **2001**, *38*, 179–189. [CrossRef]
19. Zerva, A.; Tsantopoulos, G.; Grigoroudis, E.; Arabatzis, G. Perceived citizens' satisfaction with climate change stakeholders using a multicriteria decision analysis approach. *Environ. Sci. Policy* **2018**, *82*, 60–70. [CrossRef]
20. Manolitzas, P.; Kostagiolas, P.; Grigoroudis, E.; Intas, G.; Stergiannis, P. Data on patient's satisfaction from an emergency department: Developing strategies with the Multicriteria Satisfaction Analysis. *Data Brief* **2018**, *21*, 956–961. [CrossRef]
21. Borishade, T.; Kehinde, O.; Iyiola, O.; Olokundun, M.; Ibidunni, A.; Dirisu, J.; Omotoyinbo, C. Dataset on customer experience and satisfaction in healthcare sector of Nigeria. *Data Brief* **2018**, *20*, 1850–1853. [CrossRef]
22. Drosos, D.; Tsotsolas, N.; Skordoulis, M.; Chalikias, M. Patient satisfaction analysis using a multi-criteria analysis method: The case of the NHS in Greece. *Int. J. Prod. Qual. Manag.* **2018**, *25*, 491–505. [CrossRef]
23. Fang, Y.H.; Chiu, C.M.; Wang, E.T. Understanding customers' satisfaction and repurchase intentions: An integration of IS success model, trust, and justice. *Intern. Res.* **2011**, *21*, 479–503. [CrossRef]
24. Kadlubek, M.; Grabara, J. Customers' expectations and experiences within chosen aspects of logistic customer service quality. *Intern. J. Qual. Res.* **2015**, *9*, 265–278.
25. Tsafarakis, S.; Kokotas, S.; Pantouvakis, A. A multiple criteria approach for airline passenger satisfaction measurement and service quality improvement. *J. Air Transport. Manag.* **2018**, *68*, 61–75. [CrossRef]

26. Vijayabanu, C.; Renganathan, R.; Badrinath, V.; Vijay Anand, V.; Chandrasekar, S.; Parthasaarathy, A.K.; Ganapathi Narendra Subburam, U. Customer satisfaction in aviation industry. *Int. J. Appl. Bus. Econ. Res.* **2017**, *15*, 397–405.
27. Bouranta, N.; Siskos, Y.; Tsotsolas, N. Measuring police officer and citizen satisfaction: Comparative analysis. *Policing* **2015**, *38*, 705–721. [CrossRef]
28. Aouadni, I.; Rebaï, A.; Christodoulakis, N.; Siskos, Y. Job satisfaction measurement: The multi-criteria satisfaction analysis. *Int. J. Appl. Decis. Sci.* **2014**, *7*, 190–207. [CrossRef]
29. Chen, S.C. The customer satisfaction-loyalty relation in an interactive e-service setting: The mediators. *J. Retail. Consum. Serv.* **2012**, *19*, 202–210. [CrossRef]
30. Drosos, D.; Tsotsolas, N.; Manolitzas, P. The relationship between customer satisfaction and market share: The case of mobile sector in Greece. *Int. J. Eng. Manag.* **2011**, *3*, 87–105.
31. Ostrom, A.; Iacobucci, D. Consumer trade-offs and the evaluation of services. *J. Mark.* **1995**, *59*, 17–28. [CrossRef]
32. Woodside, A.G.; Frey, L.L.; Daly, R.T. Linking service quality, customer satisfaction and behavioral intention. *J. Health Care Mark.* **1989**, *9*, 5–17.
33. Yi, Y. A critical review of consumer satisfaction. In *Review of Marketing*; Zeithaml, V.A., Ed.; American Marketing Association: Chicago, IL, USA, 1990; pp. 68–123.
34. Aghamirian, B.; Dorri, B.; Aghamirian, B. Customer knowledge management application in gaining organization's competitive advantage in electronic commerce. *J. Theor. Appl. Electron. Comm. Res.* **2015**, *10*, 63–78. [CrossRef]
35. Kotler, P. *Marketing Management: Analysis, Planning, Implementation, and Control*; Prentice Hall: Englewood Cliffs, NJ, USA, 1991.
36. O'Sullivan, D.; McCallig, J. Customer satisfaction, earnings and firm value. *Eur. J. Mark.* **2012**, *46*, 827–843. [CrossRef]
37. Balmer, J.M.T.; Chen, W. Corporate heritage brands, augmented role identity and customer satisfaction. *Eur. J. Marke.* **2017**, *51*, 1510–1521. [CrossRef]
38. Luo, X.; Homburg, C.; Wieseke, J. Customer satisfaction, analyst stock recommendations, and firm value. *J. Mark. Res.* **2010**, *47*, 1041–1058. [CrossRef]
39. O'Connell, V.; O'Sullivan, D. The impact of customer satisfaction on CEO bonuses. *J. Acad. Mark. Sci.* **2011**, *39*, 828–845. [CrossRef]
40. Ryu, K.; Lee, H.R.; Kim, W.G. The influence of the quality of the physical environment, food, and service on restaurant image, customer perceived value, customer satisfaction, and behavioral intentions. *Int. J. Contemp. Hosp. Manag.* **2012**, *24*, 200–223. [CrossRef]
41. Anderson, E.W.; Mittal, V. Strengthening the satisfaction-profit chain. *J. Serv. Res.* **2000**, *3*, 107–120. [CrossRef]
42. Alegre, J.; Cladera, M. Analysing the effect of satisfaction and previous visits on tourist intentions to return. *Eur. J. Mark.* **2009**, *43*, 670–685. [CrossRef]
43. Steven, A.B.; Dong, Y.; Dresner, M. Linkages between customer service, customer satisfaction and performance in the airline industry: Investigation of non-linearities and moderating effects. *Transp. Res. Part E* **2010**, *48*, 743–754. [CrossRef]
44. Banker, R.D.; Mashruwala, R. The moderating role of competition in the relationship between nonfinancial measures and future financial performance. *Contemp. Acc. Res.* **2007**, *24*, 763–793. [CrossRef]
45. Jani, D.; Han, H. Investigating the key factors affecting behavioral intentions: Evidence from a full-service restaurant setting. *Int. J. Contemp. Hosp. Manag.* **2011**, *23*, 1000–1018. [CrossRef]
46. Ranaweera, C.; Prabhu, J. On the relative importance of customer satisfaction and trust as determinants of customer retention and positive word of mouth. *J. Target. Meas. Anal. Mark.* **2003**, *12*, 82–90. [CrossRef]
47. Pantelidis, I.S. Electronic meal experience: A content analysis of online restaurant comments. *Cornell Hosp. Q.* **2010**, *51*, 483–491. [CrossRef]
48. Tsai, H.T.; Huang, H.C. Determinants of e-repurchase intentions: An integrative model of quadruple retention drivers. *Inf. Manag.* **2007**, *44*, 231–239. [CrossRef]
49. Williams, P.J.; Naumann, E. Customer satisfaction and business performance: A firm-level analysis. *J. Serv. Mark.* **2011**, *25*, 20–32. [CrossRef]
50. Yeung, M.C.H.; Ennew, C.T. Measuring the impact of customer satisfaction on profitability: A sectoral analysis. *J. Target. Meas. Anal. Mark.* **2001**, *10*, 106–116. [CrossRef]

51. Dong, K.Y.; Jeong, A.P. Perceived service quality: Analyzing relationships among employees, customers, and financial performance. *Int. J. Qual. Reliabil. Manag.* **2007**, *24*, 908–926.
52. Chi, C.G.; Gursoy, D. Employee satisfaction, customer satisfaction, and financial performance: An empirical examination. *Int. J. Hosp. Manag.* **2009**, *28*, 245–322. [CrossRef]
53. Gruca, T.S.; Rego, L.L. Customer satisfaction, cash flow, and shareholder value. *J. Mark.* **2005**, *69*, 115–130. [CrossRef]
54. Winkler, G.; Schwaiger, M.S. Is customer satisfaction driving revenue—A longitudinal analysis with evidence from the banking industry? *J. Bus. Econ. Res.* **2004**, *2*, 11–22. [CrossRef]
55. Grewal, R.; Chandrashekaran, M.; Citrin, A.V. Customer satisfaction heterogeneity and shareholder value. *J. Mark. Res.* **2010**, *47*, 612–626. [CrossRef]
56. Merrin, R.P.; Hoffmann, A.O.I.; Pennings, J.M.E. Customer satisfaction as a buffer against sentimental stock-price corrections. *Mark. Lett.* **2013**, *24*, 13–27. [CrossRef]
57. Tuli, K.; Bharadwaj, S. Customer Satisfaction and Stock Returns Risk. *J. Mark.* **2009**, *73*, 184–197. [CrossRef]
58. Gounaris, S.P.; Avlonitis, J.G.; Kouremenos, A.; Papavassiliou, N.; Papathastopoulou, P. Market share and customer satisfaction: What is the missing link? *J. Euromark.* **2002**, *10*, 61–82. [CrossRef]
59. Rego, L.L.; Morgan, N.A.; Fornell, C. Reexamining the Market Share—Customer Satisfaction Relationship. *J. Mark.* **2013**, *77*, 1–20. [CrossRef]
60. Rossomme, J. Customer satisfaction measurement in a business-to-business context: A conceptual framework. *J. Bus. Ind. Mark.* **2003**, *18*, 179–195. [CrossRef]
61. Raj Kumar, R.J.; Krishnaven, V. Satisfaction of industrial customers with regard to usage of zippers. *J. Adv. Res. Dyn. Control Syst.* **2017**, *9*, 144–148.
62. Gerson, R.F. *Measuring Customer Satisfaction: A Guide to Managing Quality Service*; Crisp Publications: Menlo Park, CA, USA, 1993.
63. Siskos, Y.; Grigoroudis, E.; Zopounidis, C.; Saurais, O. Measuring customer satisfaction using a collective preference disaggregation model. *J. Glob. Optim.* **1998**, *12*, 175–195. [CrossRef]
64. Chalikias, M.; Drosos, D.; Skordoulis, M.; Tsotsolas, N. Determinants of customer satisfaction in healthcare industry: The case of the Hellenic Red Cross. *Int. J. Electron. Mark. Retail.* **2016**, *7*, 311–321. [CrossRef]
65. Medjoudj, R.; Aissani, D.; Haim, K.D. Power customer satisfaction and profitability analysis using multi-criteria decision making methods. *Int. J. Electr. Power Energy Syst.* **2013**, *45*, 331–339. [CrossRef]
66. Walsh, G.; Dinnie, K.; Wiedmann, K.P. How do corporate reputation and customer satisfaction impact customer defection? A study of Private Energy Customers in Germany. *J. Serv. Mark.* **2006**, *20*, 412–420. [CrossRef]
67. Elliot, J.; Serna, C. Managing customer satisfaction involves more than improving reliability. *Electr. J.* **2005**, *18*, 84–89. [CrossRef]
68. Mutua, J.; Ngui, D.; Osiolo, H.; Aligula, E.; Gachan, J. Consumers satisfaction in the energy sector in Kenya. *Energy Policy* **2012**, *48*, 702–710. [CrossRef]
69. J.D. Power. *Gas Utility Business Customer Satisfaction Study*; J.D. Power: Troy, MI, USA, 2016.
70. J.D. Power. *Gas Utility Residential Customer Satisfaction Study*; J.D. Power: Troy, MI, USA, 2016.
71. American Customer Satisfaction Index. *ACSI Utilities, Shipping, and Health Care Report*; ACSI: Ann Arbor, MI, USA, 2017.
72. Liu, B.; Zhang, T.; Zhou, W.; Chan, X. Research of electricity customer satisfaction evaluation on service blueprint. *Wirel. Commun. Netw. Mob. Comput.* **2007**, 3168–3171.
73. Chodzaza, G.E.; Gombachika, H.S. Service quality, customer satisfaction and loyalty among industrial customers of a public electricity utility in Malawi. *Int. J. Energy Sect. Manag.* **2013**, *7*, 269–282. [CrossRef]
74. Jannadi, O.A.; Al-Saggaf, H. Measurement of quality in Saudi Arabian service industry. *Int. J. Qual. Reliabil. Manag.* **2000**, *17*, 949–966. [CrossRef]
75. Medjoudj, R.; Laifa, A.; Aissani, D. Decision making on power customer satisfaction and enterprise profitability analysis using the Analytic Hierarchy Process. *Int. J. Prod. Res.* **2012**, *50*, 4793–4805. [CrossRef]
76. Ibáñez, V.A.; Hartmann, P.; Calvo, P.Z. Antecedents of customer loyalty in residential energy markets: Service quality, satisfaction, trust and switching costs. *Serv. Ind. J.* **2006**, *26*, 633–650. [CrossRef]
77. Hartmann, P.; Ibáñez, V.A. Managing customer loyalty in liberalized residential energy markets: The impact of energy branding. *Energy Policy* **2007**, *35*, 2661–2672. [CrossRef]

78. García-Acebrón, C.; Vázquez-Casielles, R.; Iglesias, V. The effect of perceived value and switching barriers on customer price tolerance in industrial energy markets. *J. Bus.-to-Bus. Mark.* **2010**, *17*, 317–335. [CrossRef]
79. Fuels Europe. *Statistical Report 2018*; Fuels Europe: Brussels, Belgium, 2018.
80. Eurostat. Natural Gas Supply Statistics. 2019. Available online: https://ec.europa.eu/eurostat/statisticsexplained/index.php?title=Natural_gas_consumption_statistics&oldid=88292 (accessed on 3 February 2019).
81. Public Company of Gas Supply of Greece (DEPA). Commercial Operation. 2015. (In Greek)Available online: http://www.depa.gr/content/article/002003007/112.html (accessed on 20 December 2018).
82. Ipsilandis, P.G.; Samaras, G.; Mplanas, N. A multicriteria satisfaction analysis approach in the assessment of operational programmes. *Int. J. Proj. Manag.* **2008**, *26*, 601–611. [CrossRef]
83. Grigoroudis, E.; Siskos, Y. Preference disaggregation for measuring and analysing customer satisfaction: The MUSA method. *Eur. J. Oper. Res.* **2002**, *143*, 148–170. [CrossRef]
84. Ipsos; London Economics; Deloitte. *Second Consumer Market Study on the Functioning of the Retail Electricity Markets for Consumers in the EU. Country Fiches*; European Commission: Brussels, Belgium, 2016.
85. Agnihotri, R.; Gabler, C.B.; Itani, O.S.; Jaramillo, F.; Krush, M.T. Salesperson ambidexterity and customer satisfaction: Examining the role of customer demandingness, adaptive selling, and role conflict. *J. Pers. Sell. Sales Manag.* **2017**, *37*, 27–41. [CrossRef]
86. Kumar, V.; Batista, L.; Maull, R. The impact of operations performance on customer loyalty. *Serv. Sci.* **2011**, *3*, 158–171. [CrossRef]
87. Drosos, D.; Tsotsolas, N.; Chalikias, M.; Skordoulis, M.; Koniordos, M. Evaluating customer satisfaction: The case of the mobile telephony industry in Greece. In *Communications in Computer and Information Science*; Kravets, A., Shcherbakov, M., Kultsova, M., Eds.; Springer International Publishing: Cham, Switzerland, 2015; pp. 249–267.
88. Jacoby, J. Consumer research: A state-of-the-art review. *J. Mark.* **1978**, *42*, 87–96. [CrossRef]

© 2019 by the authors. Licensee MDPI, Basel, Switzerland. This article is an open access article distributed under the terms and conditions of the Creative Commons Attribution (CC BY) license (http://creativecommons.org/licenses/by/4.0/).

Article

# Addressing Energy Poverty through Transitioning to a Carbon-Free Environment

Sofia-Despoina Papadopoulou, Niki Kalaitzoglou, Maria Psarra, Sideri Lefkeli, Evangelia Karasmanaki * and Georgios Tsantopoulos

Department of Forestry and Management of the Environment and Natural Resources, Democritus University of Thrace, 68200 Orestiada, Greece; sofiapapadopoulou93@gmail.com (S.-D.P.); kalaitzoglouniki@gmail.com (N.K.); maria.psarra72@gmail.com (M.P.); roylaleu@hotmail.gr (S.L.); tsantopo@fmenr.duth.gr (G.T.)
* Correspondence: evagkara2@fmenr.duth.gr; Tel.: +306-983-605-600

Received: 14 April 2019; Accepted: 6 May 2019; Published: 8 May 2019

**Abstract:** The excessive consumption of fossil fuels not only leads to resource depletion, but also involves negative environmental effects on both public health and the climate. However, Greece's renewable energy (RE) capacity is considerable and could meet a great part of the country's energy needs while helping to tackle the ecological problem our planet faces. At the same time, the deployment of renewable energy sources (RES) can facilitate the creation of new jobs and enable households to become energy independent, while addressing energy poverty. The present study investigates the views and attitudes of citizens of the Thessaloniki conurbation towards RES. To collect the data, structured questionnaires were used, which were completed through personal interviews. Moreover, random sampling was performed to select the sample, and in total 420 citizens participated in the survey. Results showed that the respondents supported the replacement of lignite plants with renewable energy sources since they perceived that they constitute a necessary solution providing opportunities for economic growth and improvement to their quality of life. Finally, the vast majority expressed increased interest in future investment in photovoltaic systems, which in their opinion could contribute to improving air quality and increasing the energy independence not only of Greece but also of households.

**Keywords:** Renewable energy sources; energy poverty; energy transition; citizen attitudes; Thessaloniki

## 1. Introduction

Economic growth is dependent on energy which supports economic activity, enhances productivity and meets basic human needs. The energy sector lies at the core of challenges and European countries' attention has turned to renewable energy sources due to the uncertainty of fossil reserves and their negative environmental effects. To alleviate environmental problems, the member states of the European Union (EU) are required to take immediate action for the development of new energy production technologies, and to that end the EU has established a legislative framework which provides for the promotion and use of renewable energy sources [1]. In particular, the Directive 2001/77/EC was introduced, which provided that 12% of the energy produced within the EU should be generated from renewable sources [2]. A few years later, the European Commission proposed the Directive 2009/28/EC, encouraging the development of renewable energy technologies [3].

In the context of the European Energy Policy, the National Renewable Energy Action Plan was established in Greece for energy saving. According to the plan, renewable energy sources must be promoted for electricity generation and Greece is bound to meet certain environmental commitments regarding the set targets of 2020. To achieve the set targets, the suggested RES technologies involved both onshore and offshore wind energy, photovoltaic technology, as well as geothermal energy [4].

To plan effective policies for the sustainable development of natural resources, it is significant to fully understand and consider human environmental behavior [5]. In other words, it appears that knowing the attitudes and behavior of citizens is particularly important to energy policy design. Indicatively, the Swedish Energy Policy has attributed great importance to environmental behavior and was greatly affected by the Swedish citizens' attitudes towards energy production and management [6]. In addition, it was indicated that the understanding of citizens' attitudes and behavior in terms of sustainable energy management has been rapidly increasing in the past years, mainly due to the need for better communication between decision-makers and the public being highlighted [7].

In our age, the estimations of energy technology are extensive in environmental and economic terms and at the same time citizens' concerns about environmental issues and energy saving have increased. For example, research has identified that in the years 2001–2002 the inhabitants of the Greek islands were supportive of the existing wind parks in contrast to the inhabitants of the mainland who had a negative attitude [1]. More specifically, the inhabitants of the mainland expressed either divided opinions on or negative attitudes to wind energy applications, highlighting the need for providing residents with additional information on wind power [8]. As for photovoltaic system application, the research findings of Hondo and Baba [9] are of great interest since this research team discovered that 200 household heads in Lida were interested in electricity saving and estimated the costs and benefits from photovoltaic application, while as community members communicated with each other on the system's usefulness and developed a more positive behavior concerning the management of the natural environment [9].

In periods of recession and financial credit, increased development levels cannot ensure the continuous access to efficient forms of energy and consequently the phenomenon of energy poverty emerges [10]. According to the Community Directive 2009/72/EC, in Europe energy poverty is reaching worrying proportions and requires an immediate response [11]. In Greece, energy poverty stems from the economic crisis and was first noticed in the initial years of the crisis, while the scale of energy poverty compelled the Greek state to form an appropriate policy to address this issue.

To put this differently, energy poverty emerges when lower-income households have difficulty covering energy costs for electricity and heating purposes. Installations of renewable energy, and especially photovoltaic systems, provide households with the opportunity to become energy independent and combat the problem of energy poverty. A necessary condition for this to happen is that citizens invest in renewables. However, no study has examined the investment willingness and attitudes towards renewable energy of citizens residing in large cities such as Thessaloniki. Hence, the main aim of the present study is to examine the views of the citizens of the Thessaloniki conurbation on a set of issues concerning energy production from different sources, investments in renewable energy and predominantly in photovoltaic systems, which can contribute significantly to the reduction of energy poverty. The study findings can make an important contribution because they can be used as an effective tool by policymakers to make decisions on policies and adapt their decisions to the new data. To be more specific, having insights into citizens' mindset can help policymakers formulate policies and introduce measures which correspond to citizens' expectations and needs. In the long run, a favorable climate for citizen investment in RES can be fostered, which in turn will address energy poverty.

## 2. Theoretical Background

Energy, energy use and carbon emissions reduction are widely recognized as the most important environmental issues of our time, while they are also crucial to economic and social development, as well as the improvement of life quality [12,13]. To achieve a substantial reduction in greenhouse gas emissions, increasing the share of renewable energy sources in the total energy production is becoming the most important aspect of strategies in many countries [14].

Renewable energy sources play a key role in environmental protection since their exploitation does not harm the environment due to the lack of pollutants or gases which increase the risk of climate change.

Simultaneously, the use of RES for electricity production can contribute greatly not only to reducing the dependence on the expensive imported oil but also to reinforcing energy security [15,16]. Apart from the reduction in environmental pollution and the enhancement of energy saving, the creation of new jobs consists another notable advantage of renewable energy deployment [17]. Yet, investments in RES often involve external costs and benefits which should be taken into account to achieve socially optimal investments [14].

The adoption and promotion of renewable energy technologies have become more and more important to environmental and economic sustainability, as global concerns about climate change and dependence on imported fossil fuels are increasing [18,19]. Specifically, solar energy is not only abundant, but also consists one of the fastest growing renewable sources, since it does not contribute to carbon emissions and hence is not harmful to the environment [20].

At the European Union level, the pursued energy policy has attempted to develop coherent strategies to set objectives common to all countries which include pollution reduction and growth in renewable energy use. In the context of the Europe 2020 strategy, the countries of the EU are committed to maintain the 20/20/20 targets in terms of climate/energy. In essence, this means that there should be 20% reduction in greenhouse gas emissions compared to the levels recorded in 1990, 20% increase in the share of renewable sources of energy in total energy production and 20% increase in energy efficiency. Notably, these targets are inseparable from the targets which are set for education, employment, innovation and fight against poverty and at the same time they are of vital importance for the overall success [21]. The reduction in greenhouse gas emissions by 40% below the 1990 levels as well as the other targets relating to RES and energy efficiency in EU are included in the "Energy Roadmap 2050" [22]. Hence, the competitiveness and security of energy supply can be secured, while the transition to a low-carbon economy and energy system decentralization can be supported through energy technology innovation [23].

Today, the European Union constitutes the biggest economy which is legally bound to attain 27% energy consumption from RES by 2030. To achieve this objective, electricity production from RES must reach 1600–1700 TWh by 2030, which is double than the corresponding rate in 2014 [24]. According to the latest reports, the European Union has made great progress since 2005 and is well on course to meet the 2020 renewable energy targets [25]. However, research has indicated that the Directive 2009/28/EC on EU action, which is responsible for the application of the general policy on the RES targets by member states, does not provide the structure or the content needed by member states to apply the RES policy and achieve the set targets [3].

In times of austerity and increasing energy poverty, the improvement of energy autarky through RES is characterized as an economic savior. At a national level, renewable energy sources can constitute a long-term solution to international dependence on fuel imports, pan-European energy security and a way to eliminate failed economies due to financial abuse [26]. Ensuring access to imported energy resources in advantageous and competitive terms is the primary target for the international relations of the energy dependent countries. International relations are of great significance to the solution of the energy issue especially in countries lacking energy autonomy such as Greece [27].

The use of electricity is a prerequisite for modern living, however, energy production entails high monetary and environmental costs. For this reason, energy saving measures are required and energy production must be covered as much as possible by RES [15], rendering the harmonization and adjustment of the Greek energy market and institutional framework to the current international trends, perceptions and requirements [27]. It is reasonable therefore that the focus is on specific sectors. These involve the liberalization of the gas and electricity market, increase in the production of electricity from renewable energy sources, reduction in energy generated by conventional technologies, improvement of energy efficiency, energy saving and environmental protection with a focus on the domestic sector. Meanwhile, the Greek energy market is presently experiencing rapid and radical changes [28].

At this point it is interesting to provide a context of the energy situation in Greece. The photovoltaic system in Greece began in 2006 and was updated in 2011, providing potential financial stability and consisting an alternative to oil and lignite, while the latter is today still dominant in the energy sector. The program promoted new opportunities for both small-scale photovoltaic applications for households and large-scale photovoltaic parks producing enough energy for international exports. However, in such cases there is often opposition to be faced, while the energy program's success or failure can be affected by the understanding of local social relations and historical consciousness, but also by government policy [26]. In Greece, since 2009 a type of unregulated investor policy has seemed to prevail, which is characterized by a high number of applications for constructing renewable energy projects. Such projects are often implemented without strategic planning and involve large-scale construction projects in small island regions while environmental and cultural limitations specified in the relevant legislation are being violated. Meanwhile, local communities are not informed and existing industries, rural and domestic activities, as well as land qualities are neglected giving thus rise to common sense [29]. Considering the above, Greece must evaluate two more elements. First, local and regional authorities now participate in environmental policymaking and implementation without however possessing the necessary fiscal and administrative capacities while signs of customer relations and corruption are observed [30]. Second and most important, the EU involvement through troika [31] has been dramatically affecting the conditions of environmental development in Greece since the year 2010 [32] through a set of legislative changes including investment facilitation and national economy boosting [33]. It is clear that the success of national policy is inevitably directly related to international relations [34]. Possibly, the most important issue that Greece faces in terms of policies, is the constant political regression which causes alterations in the institutional and economic frameworks relating to RES development. The resulting uncertainty consists an inhibiting factor for investors and at the same time frustration occurs among social groups residing in areas where renewable energy projects are to be implemented [15,35].

In terms of the total electricity production in Greece, the official data of the Ministry of Environment and Energy are of great interest and show that there is a constant increase in electricity production. Specifically, an increasing trend in electricity production is observed and the greatest share of energy production is lignite-based, while natural gas has been introduced to the energy mix and a gradual increase in energy from renewables is noticed [36]. Interestingly, the expectations of natural gas development seem to play a preventive role in decision-making regarding the long-term development of solar photovoltaic energy [37]. Meanwhile, the most important impacts of the operation of fossil fuel power plants involve the harmful effects on human health and the increased concentration of carbon dioxide in the air [38]. Acknowledging these effects, Law No 3851/2010 provided mandatory national targets for the RES share in overall energy consumption [39], highlighting the cooperation between industry and government [18,40] which should also cooperate with households [41]. This way the benefits of microgeneration as part of a wider shift towards reducing energy demand and consumer behavior change can be secured. That is particularly important especially because households often ignore their important contribution to these targets [42]. Moreover, suppliers should endeavor to better understand the perceptions of customers, so they can develop technology products which are attractive to the current household owners [34].

In this context, many private and public investors have repeatedly made efforts to install new RES-based plants inducing often serious local opposition [40]. For this reason, the different character of local preferences for RES should be taken into consideration [4]. The dynamic of public commitment in the technological development of RES [43] is able to address public attitudes which constitute obstacles or barriers, by shaping citizens' perceptions in this direction [6,44,45]. This can be achieved by involving citizens in the initial planning and implementation stages [46], and especially citizens residing close to installation sites [47]. Additionally, it would be useful to provide citizens with information on the major advantages of wind and solar system use, which are not only effective, economical, efficient and environmentally friendly sources, but also ones that generate lower greenhouse gas emissions. However, it is possible that the manufacturing procedures of the materials of these systems cause harmful emissions as it is the case with other products. Consequently, the possibility of negative effects must be monitored within a comparative life-cycle framework [38] in rural rather than urban areas [48] to achieve a sustainable future [16,44,46,47]. It is noteworthy that renewable energy sources cannot bring about significant changes, but people must change habits so that society can attain a "green future" [15]. Public opinion is generally considered particularly inconsistent, while perceptions are referred to as social and personal [49] and attitudes differ substantially from one project to another, among regions but also within the same region and can change over time. In reality, public opinion depends on various factors such as ignorance, misinformation regarding the advantages of a specific technology, local environmental conditions, national energy policies, residents' experience in such projects, renewable energy technology costs [40,44,45]. Consumer behavior is complicated and rarely follows traditional economic decision-making theories. When people buy goods, they often believe that they take smart decisions and act "rationally". Yet, in their everyday life people deviate from the model of "rational behavior" [50], according to which a person estimates the cost and advantages of alternative solutions objectively before making the final decision. Even consumers with strong material motives possessing adequate knowledge and motives to act in sustainable ways are likely to change their behavior towards a more desirable direction [50]. However, high knowledge of the basic principles and contribution of RES is observed, while great awareness levels about environmental problems are recorded. More specifically, there is substantial support for RES installations [34,51], especially for existing RES projects (particularly photovoltaic systems, wind parks and hydropower applications) without rejecting new installations which does not resonate with the NIMBY phenomenon [1,40,46,49,52]. Awareness about the issue of climate change and other environmental problems as well as high concern levels about the environment regarding the future seems to be a factor validating renewable energy [6,19,51]. Simultaneously, most expressed concerns are based on the impact of photovoltaics on land-use and of hydropower stations on flora and fauna and not so much on the noise produced by wind turbines [18,40]. For example, in regions of Greece with high wind capacity and investment interest the public has a supportive attitude towards wind energy applications including both existing and new turbines, whereas in continental Greece public opinion is divided [8]. Nevertheless, the same study revealed that a minority consisting mainly of farmers was acutely opposite to wind energy applications ignoring the economic benefits [8].

Moreover, a possible existence of the snowball effect was noticed regarding the acceptance and support for RES if some minorities changed their lifestyle [2], wishing that the responsibility of climate change action is transferred to governments. However, the cooperation ability of citizens (collective action) on a moral and voluntary basis to achieve behavioral change was deemed limited. This has consequences for policies on energy and resources usage. In this context, it may be useful that policymakers first establish policies which are considered most appropriate for the public and second take initiatives through information and communication campaigns examining citizens' concerns about the usefulness of government approaches [5].

On the other side, citizens' limited sense of personal responsibility suggests that energy efficiency and RES installation on individual level is pointless and lacks significant motives. Shares of citizens have expressed many arguments regarding the safeguarding of future generations and lifestyle changes to justify RES installations. In addition, there are positive correlations between energy efficiency and lifestyle changes, but the latter are not fully negotiable [1,5,53]. A recent study has identified the willingness of citizen majority to pay more for electricity generated by RES, either in the form of payment for public projects or domestic system installation which provide long-term economic profit [6,15]. What is more, another study has indicated that certain groups of citizens were more willing to pay for renewables than others. These involved high earners, owners of large houses, individuals with high knowledge and awareness levels about energy and climate change, as well as people facing electricity shortage [54]. Apart from citizen groups, tourists also exhibited positive attitudes to RES since they stated their preference for hotels which have invested in energy saving measures and RES even if they were required to pay higher prices [55].

Yet, the cost still constitutes a notable barrier to the adoption of pro-environmental behavior in the form of adopting renewable energy systems, while the availability and price of technologies are constantly changing. In relation to solar energy, it has been indicated that not owning a roof or a plot of land consists a significant hindrance to photovoltaic (PV) installation [35,56] and, at the same time, plans to move to another house and lack of house ownership affect consumer perception [57]. On the other side, obstacles to wind energy implementation often occur when the local public resists installations on grounds of the visual impact of such projects on the landscape [58–60]. It therefore becomes clear that RES policymaking should ensure that the long-term benefits of RES investment will offset the cost of participating in positive environmental behavior through "green subsidies" [61].

From the citizens' viewpoint, an overall support for RES development was observed, which was closely related to the dissemination of solar water heaters [53]. This support can be accounted for by the fact that a significant part of the household's warm water demand can be covered by solar water heating systems which cover even 70% while involving a short depreciation time [16]. Simultaneously, it would be beneficial to inform citizens about topics on photovoltaic system installation including cost reduction due to system integration into building parts such as windows and tents [62]. Moreover, it has been found that households which are familiar with photovoltaic systems tend to enhance their daily environmental behavior [9]. However, the same does not apply to wind energy since citizens are prevented from adopting it due to turbines' high cost and size granting thus exclusivity to multinational companies [31,52].

The understanding of citizens' attitudes and the stages undergone before deciding to adopt RES is not only varied but can also have great influence. Hence, the understanding can enable those in governments to design more effectively strategic policies which aim at enhancing citizen acceptance of RES while acknowledging citizens' expectations which mainly involve financial incentives. Moreover, they will be able to reinforce the appropriate supportive and educational means. Second, greater control of marketing strategy should be exercised to make this technology more attractive and affordable even for households suffering from or threatened by energy poverty, enabling investments in RES-based electricity production.

## 3. Materials and Methods

The present study was performed in the city of Thessaloniki. To achieve the research objectives, it was deemed necessary to use structured questionnaire and personal interviews because these are considered effective when it is attempted to capture the views of a large number of respondents [63]. In particular, the questionnaire allows the participation of a great number of participants facilitating the identification of possible differentiated or opposite tendencies regarding electricity production from lignite and RES, which would not have possibly been discovered by a different quantitative method. Hence, this method contributes more effectively to a more thorough understanding of the topics investigated by this study.

To design the questionnaire, the relevant literature on citizen views and attitudes towards household photovoltaic system installation was taken into account [9,15,16,34,40,42,46]. To analyze the collected data, descriptive statistics, the non-parametric Friedman test and factor analysis were the chosen methods. The Friedman test is a statistical method that is applied to compare the values of three or more groups of variables which are correlated. Also, the distribution of the Friedman test is $\chi^2$ distribution with degrees of freedom $(df)$ $df = k - 1$, where $k$ stands for the number of teams or samples. In addition, the test analyses the values of variables for each subject separately and estimates the mean ranks of classification values for each variable [64,65]. On the other side, factor analysis is a statistical method which investigates whether there are common factors within a group of variables. More specifically, principal component analysis was performed here, which is on the basis of the spectral analysis of the variance (or correlation) matrix [66,67]. Moreover, the criterion employed for the significance of the principal components was the one suggested by Guttman and Kaiser, which states that the limit for acquiring the required number of principal components is determined by the eigenvalues which are equivalent to or greater than one. Then, we carried out a matrix rotation of the principal components using the Kaiser's varimax rotation method to obtain more coherent results [68].

Regarding the sampling method, we selected the simple random sampling because it is not only simple but also requires less knowledge about the population than any other sampling method. In total, 420 residents in the Thessaloniki conurbation completed the questionnaire and the data collection was implemented in the period between June–July 2018.

## 4. Results

Regarding the socio-demographic characteristics of the participants, female respondents (53.3%) outnumbered their male counterparts, while more than half (56%) were aged between 31 and 50 years and considerably fewer were aged between 51 and 60 years. In terms of educational level, it was indicated that overall the participants were highly educated, since over half of the surveyed citizens were university or technical institution graduates, whereas only one in five respondents had attended merely high school. Moreover, in terms of occupation, most participants were employed in the private sector (30.7%), whereas 20.5% were public servants and 12.6% were unemployed. In addition, 11.7% were higher education students, while the percentages of freelancers, farmers and people involved in housework were particularly low. Finally, regarding the gross annual household income, 21.2% of respondents earned from 10.001 to 20.000€, while 19% received from 5001 to 10,000€ and 17.9% less than 5000€. At the same time, 15.5% of participants received from 20,001 to 30,000€ and only 6% more than 30,000€. It is also noteworthy that a substantial proportion (20.5%) of the surveyed citizens did not wish to report their income.

Citizens were first asked whether they agreed that renewable energy sources offer opportunities for economic growth. As presented in Table 1, most participants, by 84.8%, agreed with this statement, whereas only 5.5% disagreed. Moreover, as it can be seen in Table 2 more than half respondents (53.5%) considered that Greece has not invested in renewable energy sources, while only 11.4% perceived that the Greek state has invested in RES.

**Table 1.** Frequency and percentages regarding the degree of agreement with the view that renewable energy sources offer opportunities for economic growth.

|  | Frequency | Percentage (%) |
|---|---|---|
| Do not know | 12 | 2.9 |
| Totally disagree | 10 | 2.4 |
| Disagree | 13 | 3.1 |
| Neither agree nor disagree | 29 | 6.9 |
| Agree | 191 | 45.5 |
| Totally agree | 165 | 39.3 |
| Total | 420 | 100.0 |

**Table 2.** Frequency and percentages concerning participants level of agreement with the view that the Greek state has invested in renewable energy sources.

|  | Frequency | Percentage (%) |
|---|---|---|
| Do not know | 65 | 15.5 |
| Totally disagree | 61 | 14.5 |
| Disagree | 164 | 39.0 |
| Neither agree or disagree | 82 | 19.5 |
| Agree | 27 | 6.4 |
| Totally agree | 21 | 5.0 |
| Total | 420 | 100.0 |

Next, the citizens' degree of agreement with the gradual reduction in lignite-based power generation and transition to environmentally friendlier energy types was investigated. As Table 3 shows, the vast majority (90.7%) supported this change and only 3.1% of respondents disagreed, while 6.2% neither agreed nor disagreed. Afterwards, citizens' view on the construction of new lignite plants in regions where lignite reserves are located was examined. Remarkably, most citizens were opposed to constructing new lignite plants (58.8%), while those supporting the construction were 16.9% and those taking a neutral position were 24.3% (Table 4).

Table 3. Frequency and percentages concerning respondents' view on the gradual reduction in lignite-based electricity production and transition to environmentally friendly energy types.

|  | Frequency | Percentage (%) |
| --- | --- | --- |
| Totally disagree | 6 | 1.4 |
| Disagree | 7 | 1.7 |
| Neither agree nor disagree | 26 | 6.2 |
| Agree | 158 | 37.6 |
| Totally agree | 223 | 53.1 |
| Total | 420 | 100.0 |

Table 4. Frequency and percentages relating to citizens' degree of agreement with the construction of new lignite plants in areas with lignite resources.

|  | Frequency | Percentage (%) |
| --- | --- | --- |
| Totally disagree | 103 | 24.5 |
| Disagree | 144 | 34.3 |
| Neither agree nor disagree | 102 | 24.3 |
| Agree | 61 | 14.5 |
| Totally agree | 10 | 2.4 |
| Total | 420 | 100.0 |

Afterwards, the participants evaluated the impacts which would arise if the existing lignite plants were replaced with installations of renewable energy sources. To scrutinize the data, responses were ranked using the non-parametric Friedman test (Table 5). According to the test results, the surveyed citizens of the Thessaloniki conurbation regarded the improvement of life quality in the nearby areas as the most important advantage if renewable energy sources replaced exiting lignite plants (mean rank 5.93). This was followed by the protection of the regional flora and fauna (mean ranks of 5.86 and 5.84, respectively). However, tourism development received the last ranking since only few respondents perceived that RES transition would contribute to the development of tourism in the surrounding areas.

Table 5. The rankings of the non-parametric Friedman test regarding citizens' opinion on the impacts of replacing lignite plants with renewable energy sources.

| Impacts | Mean Rank |
| --- | --- |
| Improved life quality | 5.93 |
| Regional economic development | 4.48 |
| Creation of new jobs | 4.34 |
| Tourism development | 3.83 |
| Landscape enhancement | 5.29 |
| Local fauna protection | 5.84 |
| Local flora protection | 5.86 |
| Agriculture development | 4.71 |
| Livestock farming development | 4.71 |
| $n = 420$ Chi-Square = 388.294 $df = 8$ $p < 0.001$. | |

To gain further insights into participants' views on impacts of lignite replacement with RES, factor analysis was deemed appropriate. Before proceeding, the eligibility of the data had to be tested. Hence, the Cronbach's alpha value was 0.947, the Keiser-Meyer Olkin index value was 0.852 and Bartlett's test of sphericity gave $\chi^2 = 7782.354$, $df = 36$ and $p < 0.001$, indicating our data were suitable for factor analysis. After performing Varimax rotation, two factors emerged for the multivariate "Impacts of replacing lignite plants with renewable energy sources on the surrounding areas" (Table 6). Specifically, the first factor (PC1) includes the variables "Local flora protection", "Local fauna protection", "Agriculture development", "Livestock farming development ", "Landscape enhancement" and "Tourism development". The second factor (PC2) contains the variables "Regional economic development", "Creation of new jobs" and "Improved life quality".

Table 6. Rotated component matrix for citizens' views on the regional impacts due to replacing lignite plants with RES.

| Variables | Rotated Matrix | |
|---|---|---|
| | PC 1 | PC 2 |
| Local flora protection | 0.923 | 0.249 |
| Local fauna protection | 0.920 | 0.260 |
| Agriculture development | 0.907 | 0.267 |
| Livestock farming development | 0.907 | 0.267 |
| Landscape enhancement | 0.900 | 0.274 |
| Tourism development | 0.774 | 0.248 |
| Regional economic development | 0.261 | 0.934 |
| Creation of new jobs | 0.180 | 0.925 |
| Improved life quality | 0.541 | 0.662 |

Then, respondents assessed different energy sources based on which they wished to be developed in Greece. At this point, responses were ranked with the non-parametric Friedman test. As tabulated in Table 7, solar energy was ranked first with a mean rank of 7.66, followed by wind energy and hydropower with mean ranks of 7.30 and 6.95, respectively. Finally, coal combustion was the least preferred energy option (mean rank 2.67).

Table 7. The application of the Friedman test for ranking respondents' evaluation of different energy production technologies.

| Energy Production Technologies | Mean Rank |
|---|---|
| Lignite combustion | 2.86 |
| Coal combustion | 2.67 |
| Oil combustion | 3.03 |
| Natural gas combustion | 5.45 |
| Hydropower | 6.95 |
| Wind energy | 7.30 |
| Solar energy | 7.66 |
| Nuclear fuels | 2.91 |
| Biofuels | 6.16 |
| $n = 420$ Chi-Square $= 2341.566$ $df = 8$ $p < 0.001$. | |

Before performing factor analysis, the suitability of the data had to be verified and thus the Bartlett test of sphericity, Cronbach Alpha and the Keiser-Meyer-Olkin measure were applied. Thus, the Cronbach's Alpha scored 0.773, the KMO index was 0.775 and Bartlett's test of sphericity gave $\chi^2 = 1981,398$, $df = 36$, $p < 0.001$. After performing the varimax rotation, two factors were extracted for the value variables. As it appears in Table 8, the variables "Solar energy", "Wind energy", "Hydropower" and "Biofuels" fell under the first factor (PC1), whereas the variables "Coal combustion", "Lignite combustion", "Oil combustion", "Nuclear fuels" and "Natural gas combustion" fell under the second factor (PC2).

Table 8. Rotated component matrix for respondents' evaluation of energy production technologies.

| Variables | Rotated Matrix | |
|---|---|---|
| | PC 1 | PC 2 |
| Solar energy | 0.910 | 0.025 |
| Wind energy | 0.910 | −0.001 |
| Hydropower | 0.874 | 0.135 |
| Biofuels | 0.612 | 0.004 |
| Coal combustion | −0.038 | 0.900 |
| Lignite combustion | −0.024 | 0.875 |
| Oil combustion | 0.092 | 0.810 |
| Nuclear fuels | 0.076 | 0.568 |
| Natural gas combustion | 0.460 | 0.517 |

The surveyed citizens were then asked whether they were interested in investing in a photovoltaic system as house owners. Remarkably, the clear majority of participants (by 91.9%) showed interest in a future investment, while only 8.1% were not interested (Table 9).

Table 9. Frequency and percentages regarding citizens' interest in investing in photovoltaic systems in the future.

| | Frequency | Percentage (%) |
|---|---|---|
| Yes | 386 | 91.9 |
| No | 34 | 8.1 |
| Total | 420 | 100.0 |

Respondents also evaluated various reasons for installing photovoltaic systems. To investigate if there were any statistical differences among the reasons, the non-parametric Friedman test was applied. As Table 10 shows, the citizens under study considered the improved air quality as the most important reason to install photovoltaic systems with a mean rank of 9.31, whereas they assigned the lowest ranking to the minimum amount of work that is often required for the system installation.

Table 10. The application of the Friedman test for ranking reasons for installing photovoltaic systems.

| Reasons for Installing Photovoltaic Systems | Mean Rank |
|---|---|
| Subsidies for the purchase of RE system | 6.47 |
| Subsidies for the maintenance of the system | 5.98 |
| Fixed and guaranteed income | 6.82 |
| Minimum amount of work | 4.75 |
| Lower-risk investment for savings | 6.05 |
| Higher profitability compared to other investments | 6.18 |
| Tax exemptions due to installation cost of RE | 6.97 |
| Tax exemptions due to maintenance cost of RE | 6.75 |
| New job positions-unemployment reduction | 7.37 |
| Enhanced social prestige-entrepreneurial activity | 5.93 |
| Reduction in pollution | 9.27 |
| Improved air quality | 9.31 |
| Increased energy independence of our country | 9.14 |
| $n = 420$ Chi-Square $= 1038.803$ $df = 12$ $p < 0.001$ | |

Before conducting factor analysis to extract factors, the adequacy of the data had to be tested. To this end, the Cronbach test of reliability, the KMO index and the Bartlett test were employed. Thus, the Cronbach's alpha value was as high as 0.885, the Keiser-Meyer-Olkin index value was 0.835 and the Bartlett test of sphericity value was 3222.800, with degrees of freedom 78 and with $p < 0.001$, confirming our data's suitability. Next, principal component analysis with varimax rotation was performed and three factors were loaded for the (Table 11). The variables "Higher profitability compared to other investments", "Enhanced social prestige-entrepreneurial activity", "Lower-risk investment for savings", "New job positions-unemployment reduction", "Tax exemptions due to installation cost of RE" and "Tax exemptions due to maintenance cost of RE" fell under the first factor (PC1). The second factor (PC2) contained the variables "Improved air quality", "Reduction in pollution" and "Increased energy independence of our country". Finally, the variables "Subsidies for the maintenance of the system", "Minimum amount of work", "Subsidies for the purchase of RE system" and "Fixed and guaranteed income" loaded on the third factor (PC3).

Table 11. Reasons for installing photovoltaic systems.

| Variable | Rotated Matrix | | |
| --- | --- | --- | --- |
| | PC 1 | PC 2 | PC 3 |
| Higher profitability compared to other investments | 0.800 | 0.084 | 0.250 |
| Enhanced social prestige-entrepreneurial activity | 0.775 | 0.216 | 0.010 |
| Lower-risk investment for savings | 0.708 | 0.186 | 0.284 |
| New job positions-unemployment reduction | 0.637 | 0.418 | 0.171 |
| Tax exemptions due to installation cost of RE | 0.595 | 0.124 | 0.495 |
| Tax exemptions due to maintenance cost of RE | 0.566 | 0.102 | 0.509 |
| Improved air quality | 0.171 | 0.921 | 0.112 |
| Reduction in pollution | 0.194 | 0.902 | 0.127 |
| Increased energy independence of our country | 0.213 | 0.881 | 0.121 |
| Subsidies for the maintenance of the system | 0.104 | 0.112 | 0.813 |
| Minimum amount of work | 0.081 | -0.016 | 0.692 |
| Subsidies for the purchase of RE system | 0.327 | 0.189 | 0.665 |
| Fixed and guaranteed income | 0.262 | 0.181 | 0.650 |

## 5. Discussion

Exceeding our expectation, participants expressed a favorable stance to environmentally benign energy sources and showed an intention to transition to a "clean" energy system, while they disagreed with the construction of new lignite plants in areas of Greece where lignite reserves are located. In disagreeing with the establishment of new lignite plants, it is possible that participants acknowledged the negative effects and approaching depletion of fossil fuels [15,35].

What is more, the respondents attached greater importance to the environmental protection rather than economic prospects of renewable energy investments. Hence, it is possible that citizens, who were concerned about the environment and its problems, would adopt renewable energy in an effort to contribute to the solution of environmental problems. The interpretation that increased environmental awareness motivates individuals to support renewable energy resonates with the research of Viklund [6], Altuntaş and Turan [19] and Ektör-Akyazi et al. [51], who have also observed that individuals with environmental awareness tend to adopt environmentally friendly behavior including the decision-making on energy choices. From this perspective, it can be seen that raising environmental awareness among social groups can play a critical role in tackling environmental issues. It is also important to note that apart from our respondents in the Thessaloniki conurbation, other studies have also indicated that citizens in other regions of Greece were supportive towards renewable energy deployment [8]. Hence, it appears that citizens throughout Greece are becoming mindful about energy and express positive attitudes, giving a sense of hope and optimism about the future of renewable energy in Greece.

Another point that is worthwhile to discuss is that respondents appeared to value the ability of renewable energy to achieve energy independence. Based on this, it is also possible that they acknowledged the potential of renewables to protect households from energy poverty. At the same time, of all renewable types citizens were more supportive towards solar energy and this support could be attributed simply to the fact that as a Mediterranean country Greece has abundant sunlight throughout the year [69] and hence it only made sense to respondents that solar energy can make the greatest contribution to increasing the national energy independence while decreasing the dependence on imported fuels.

On the other side, there are challenges to be faced for an increased installation of domestic photovoltaic systems. Most notably, the lack of incentives and accurate public information inhibit citizen investments. Regarding incentives, these should focus more on the rationalization of the guaranteed kilowatt rate and on the enhancement of the subsidies provided for green innovations, since it has been shown that the provision of incentives to household owners could result in increased building energy efficiency but also in substantial behavior change [70]. In turn, increased investment in domestic photovoltaics can effectively tackle energy poverty which is threatening many Greek households due to the economic crisis.

Much to our surprise, most respondents considered that the Greek state has not invested in renewable energy. However, Greece has already met the set targets regarding the increase in the share of renewable energy in total energy production [35]. To clarify why participants held this wrong view, a future study should investigate the citizens' level of knowledge on renewable energy implementation as well as explore the information sources from which they obtained such information.

Remarkably, the overwhelming majority of citizens (91.9%) were willing to make investments in photovoltaic systems in the future confirming previous research results [35,55], which also indicated that citizens were expressing a pronounced willingness to invest in renewable energy systems or pay more for energy generated from renewables.

Although the present study has identified positive attitudes to renewable energy, greater levels are needed to achieve the desired transition to a low-carbon society. To that end, information campaigns should be frequently held to inform citizens of all ages about the benefits of renewable energy and other environmental topics. Simultaneously, it is of paramount importance to shape positive attitudes in young individuals and this could be managed through environmental education programs targeted to school students. However, at present the study program in Greek schools is rather strict [71] preventing educators from organizing frequent environmental programs. Consequently, there are limited opportunities to provide information and raise awareness about the environment among adolescents who are the future citizens [72]. In this regard, it is considered essential to integrate courses focusing on environmental and energy topics in the academic programs of primary and secondary education. That is because it has been reported that environmental educational programs can shape pro-environmental behavior [13,73]. In turn, the adoption of environmentally conscious behavior that is shaped through educational programs aiming to create "green" consciousness in students, contributes significantly to energy saving, and a reduction in greenhouse gas emissions [55,74].

## 6. Conclusions

The present study investigated the investment willingness and the views on energy-related topics of citizens living in the Thessaloniki conurbation in order to examine whether it is possible to address energy poverty through investments in renewable energy. In view of the study's findings, it can be inferred that most participants were willing to make investments in renewables in the future and expressed positive attitudes to renewable energy. The findings highlight that citizens' positive attitudes to RES investment are particularly important, mainly because citizen investment could protect households, especially lower-income ones, from energy poverty and price fluctuations in the energy market. Moreover, higher citizen investment in renewable energy can contribute to the desired transition to a low-carbon energy system and the protection of the climate.

It is crucial to develop policies and incentives which are tailored for citizens with low incomes as such measures could not only speed up the energy transition to an environmentally harmless system, but also protect households from energy poverty. In other words, to induce citizens to invest, it is necessary to develop policies and incentives which create a favorable investment environment not only for household owners seeking to enhance their income, but also for low-income households, since the latter run a greater danger of encountering energy poverty. The major themes of these policies should be the formation of a favorable environment for RES installation enabling lower-income households to have their own energy production and consumption through low-interest lending including lower interest rates. These measures could achieve a significant reduction in the electricity bill costs for heating/cooling/domestic warm water. Moreover, lower-income households could be offered specific subsidies and tax breaks when purchasing or maintaining a renewable energy system, whereas citizens who do not own a house or a plot of land should also be provided with the opportunity to invest in renewables by renting roofs or plots of land at a low price to install photovoltaic systems.

As already mentioned, the citizens under scrutiny appeared to be highly aware of the environment and its issues. This conclusion rests upon two study findings. Specifically, in evaluating impacts resulting from the replacement of lignite plants with renewable energy sources, respondents attached greater importance to impacts that had to do with the protection of the environment, such as improved air quality and local flora/fauna protection. In addition, most participants would invest in photovoltaic systems for environmental (such as pollution reduction, air quality) rather than financial reasons (such as subsidies). Hence, the present study findings highlight that individuals with pro-environmental attitudes are more likely to make investments and in view of this inference it is of the utmost importance that strategies are developed to raise environmental awareness among different social groups.

In addition, citizens had a good grasp of different energy sources and were able to distinguish between renewable and non-renewable types. This was indicated by the results of factor analysis according to which explicitly renewable energy types fell under the first factor and conventional types under the second. Therefore, these factor loadings made apparent that citizens could identify between sustainable and unsustainable energy types.

Finally, in view of the findings certain additional recommendations could be made. In specific, similar studies should be carried out more frequently and in a systematic manner to analyze public attitudes to renewable energy and investments. That is because frequent measurements could prevent possible public disappointments or low participation in renewable energy investments and at the same time help policymakers to make timely improvements and modifications in the existing policies and incentives. In other words, the overall aims of such analyses should be to detect weaknesses and problems which affect citizens' willingness to invest in renewable energy. Moreover, further information about individuals' desires, expectations and needs in relation to RES investment should be acquired since this knowledge can enable policymakers to design policies which correspond to the current circumstances and induce citizens to make investments. From this perspective, study findings, such as the ones presented in this paper, may form the basis for a set of preparatory actions and policymaking which aim at planning properly the development and installation of renewable energy, as well as addressing energy poverty.

**Author Contributions:** M.P., N.K. and S.L. reviewed the relevant literature. S.-D.P. collected the data and performed statistical analysis. E.K. prepared the methodology. S.-D.P. and E.K. wrote the original draft and then E.K. and G.T. reviewed and edited the manuscript. G.T. supervised and coordinated the work.

**Funding:** This research received no external funding.

**Conflicts of Interest:** The authors declare no conflict of interest.

## References

1. Kaldellis, J.K.; Kapsali, M.; Katsanou, E. Renewable energy applications in Greece—What is the public attitude? *Energy Policy* **2012**, *42*, 37–48. [CrossRef]
2. West, J.; Bailey, I.; Winter, M. Renewable energy policy and public perceptions of renewable energy: A cultural theory approach. *Energy Policy* **2010**, *38*, 5739–5748. [CrossRef]
3. Michalena, E.; Hills, J.M. Renewable energy issues and implementation of European energy policy: The missing generation? *Energy Policy* **2012**, *45*, 201–216. [CrossRef]

4. Kontogianni, A.; Tourkolias, C.; Skourtos, M. Renewables portfolio, individual preferences and social values towards RES technologies. *Energy Policy* **2013**, *55*, 467–476. [CrossRef]
5. Fischer, A.; Peters, V.; Vávra, J.; Neebe, M.; Megyesi, B. Energy use, climate change and folk psychology: Does sustainability have a chance? Results from a qualitative study in five European countries. *Glob. Environ. Chang.* **2011**, *21*, 1025–1034. [CrossRef]
6. Viklund, M. Energy policy options—From the perspective of public attitudes and risk perceptions. *Energy Policy* **2004**, *32*, 1159–1171. [CrossRef]
7. Owens, S.; Driffill, L. How to change attitudes and behaviours in the context of energy. *Energy Policy* **2008**, *36*, 4412–4418. [CrossRef]
8. Kaldellis, J.K. Social attitude towards wind energy applications in Greece. *Energy Policy* **2005**, *33*, 595–602. [CrossRef]
9. Hondo, H.; Baba, K. Socio-psychological impacts of the introduction of energy technologies: Change in environmental behavior of households with photovoltaic systems. *Appl. Energy* **2010**, *87*, 229–235. [CrossRef]
10. Tomprou, Z. *Energy Poverty in Europe and Greece during Economic Crisis, Case Study: Attica Regiontle*; National Technical University: Athens, Greece, 2017.
11. Aligizaki, A. *Internal Energy Market as an "Antidote" for Energy Insecurity in Europe*; University of Piraeus, Department of International and European Studies: Piraeus, Greece, 2015.
12. Ntona, E.; Arabatzis, G.; Kyriakopoulos, G.L. Energy saving: Views and attitudes of students in secondary education. *Renew. Sustain. Energy Rev.* **2015**, *46*, 1–15. [CrossRef]
13. Lee, L.-S.; Lin, K.-Y.; Guu, Y.-H.; Chang, L.-T.; Lai, C.-C. The effect of hands-on 'energy-saving house' learning activities on elementary school students' knowledge, attitudes, and behavior regarding energy saving and carbon-emissions reduction. *Environ. Educ. Res.* **2013**, *19*, 620–638. [CrossRef]
14. Bergmann, A.; Hanley, N.; Wright, R. Valuing the attributes of renewable energy investments. *Energy Policy* **2006**, *34*, 1004–1014. [CrossRef]
15. Tampakis, S.; Arabatzis, G.; Tsantopoulos, G.; Rerras, I. Citizens' views on electricity use, savings and production from renewable energy sources: A case study from a Greek island. *Renew. Sustain. Energy Rev.* **2017**, *79*, 39–49. [CrossRef]
16. Kalogirou, S. Thermal performance, economic and environmental life cycle analysis of thermosiphon solar water heaters. *Sol. Energy* **2009**, *83*, 39–48. [CrossRef]
17. Diakoulaki, D.; Zervos, A.; Sarafidis, J.; Mirasgedis, S. Cost benefit analysis for solar water heating systems. *Energy Convers. Manag.* **2001**, *42*, 1727–1739. [CrossRef]
18. Dimitropoulos, A.; Kontoleon, A. Assessing the determinants of local acceptability of wind-farm investment: A choice experiment in the Greek Aegean Islands. *Energy Policy* **2009**, *37*, 1842–1854. [CrossRef]
19. Çakirlar Altuntaş, E.; Turan, S.L. Awareness of secondary school students about renewable energy sources. *Renew. Energy* **2018**, *116*, 741–748. [CrossRef]
20. Rustemli, S.; Dincer, F.; Unal, E.; Karaaslan, M.; Sabah, C. The analysis on sun tracking and cooling systems for photovoltaic panels. *Renew. Sustain. Energy Rev.* **2013**, *22*, 598–603. [CrossRef]
21. European Commission. Communication from the Commission, Europe 2020: A Strategy for Smart, Sustainable and Inclusive Growth. Available online: http://ec.europa.eu/europe2020/index_en.htm (accessed on 10 March 2019).
22. European Commission. Energy Roadmap. Available online: https://ec.europa.eu/energy/sites/ener/files/documents/2012_energy_roadmap_2050_en_0.pdf (accessed on 10 March 2019).
23. Neagu, O.; Teodoru, M. The Relationship between Economic Complexity, Energy Consumption Structure and Greenhouse Gas Emission: Heterogeneous Panel Evidence from the EU Countries. *Sustainability* **2019**, *11*, 497. [CrossRef]
24. Lacal Arantegui, R.; Jäger-Waldau, A. Photovoltaics and wind status in the European Union after the Paris Agreement. *Renew. Sustain. Energy Rev.* **2018**, *81*, 2460–2471. [CrossRef]
25. Scarlat, N.; Dallemand, J.-F.; Monforti-Ferrario, F.; Banja, M.; Motola, V. Renewable energy policy framework and bioenergy contribution in the European Union—An overview from National Renewable Energy Action Plans and Progress Reports. *Renew. Sustain. Energy Rev.* **2015**, *51*, 969–985. [CrossRef]
26. Knight, D.M.; Bell, S. Pandora's box: Photovoltaic energy and economic crisis in Greece. *J. Renew. Sustain. Energy* **2013**, *5*, 033110. [CrossRef]

27. Ministry of Environment and Energy. Energy Policy-International Relations. Available online: http://www.ypeka.gr/Default.aspx?tabid=275&language=el-RG (accessed on 9 March 2019).
28. Ministry of Environment and Energy Energy. Available online: http://www.ypeka.gr/Default.aspx?tabid=225&language=el-RG (accessed on 9 March 2019).
29. Katsaprakakis, D.A.; Christakis, D.G. The exploitation of electricity production projects from Renewable Energy Sources for the social and economic development of remote communities. The case of Greece: An example to avoid. *Renew. Sustain. Energy Rev.* **2016**, *54*, 341–349. [CrossRef]
30. Koutalakis, C. Chapter 9 Environmental policy in Greece reloaded: Plurality, participation and the Sirens of neo-centralism. In *Sustainable Politics and the Crisis of the Peripheries: Ireland and Greece*; Leonard, L., Botetzagias, I., Eds.; Emerald Group Publishing Limited: Bingley, UK, 2011; pp. 181–200.
31. Argenti, N.; Knight, D.M. Sun, wind, and the rebirth of extractive economies: Renewable energy investment and metanarratives of crisis in Greece. *J. R. Anthropol. Inst.* **2015**, *21*, 781–802. [CrossRef]
32. Markantonatou, M. *Diagnosis, Treatment, and Effects of the Crisis in Greece: A "Special Case" or a "Test Case"?* Max Planck Institute for the Study of Societies: Cologne, Germany, 2013.
33. Lekakis, J.N.; Kousis, M. Economic Crisis, Troika and the Environment in Greece. *South Eur. Soc. Politics* **2013**, *18*, 305–331. [CrossRef]
34. Faiers, A.; Neame, C. Consumer attitudes towards domestic solar power systems. *Energy Policy* **2006**, *34*, 1797–1806. [CrossRef]
35. Tsantopoulos, G.; Arabatzis, G.; Tampakis, S. Public attitudes towards photovoltaic developments: Case study from Greece. *Energy Policy* **2014**, *71*, 94–106. [CrossRef]
36. Ministry of Environment and Energy. Energy—Power Generation. Available online: http://www.ypeka.gr/Default.aspx?tabid=277&language=el-GR# (accessed on 9 March 2019).
37. Griffiths, S. Strategic considerations for deployment of solar photovoltaics in the Middle East and North Africa. *Energy Strateg. Rev.* **2013**, *2*, 125–131. [CrossRef]
38. Fthenakis, V. Sustainability of photovoltaics: The case for thin-film solar cells. *Renew. Sustain. Energy Rev.* **2009**, *13*, 2746–2750. [CrossRef]
39. Ministry of Environment and Energy. Renewable Energy Sources. Available online: http://www.ypeka.gr/Default.aspx?tabid=285&language=el-GR (accessed on 9 March 2019).
40. Kaldellis, J.K.; Kapsali, M.; Kaldelli, E.; Katsanou, E. Comparing recent views of public attitude on wind energy, photovoltaic and small hydro applications. *Renew. Energy* **2013**, *52*, 197–208. [CrossRef]
41. Korjonen-Kuusipuro, K.; Hujala, M.; Pätäri, S.; Bergman, J.-P.; Olkkonen, L. The emergence and diffusion of grassroots energy innovations: Building an interdisciplinary approach. *J. Clean. Prod.* **2017**, *140*, 1156–1164. [CrossRef]
42. Keirstead, J. Behavioural responses to photovoltaic systems in the UK domestic sector. *Energy Policy* **2007**, *35*, 4128–4141. [CrossRef]
43. Ozden, T.; Akinoglu, B.G.; Turan, R. Long term outdoor performances of three different on-grid PV arrays in central Anatolia—An extended analysis. *Renew. Energy* **2017**, *101*, 182–195. [CrossRef]
44. Klick, H.; Smith, E.R.A.N. Public understanding of and support for wind power in the United States. *Renew. Energy* **2010**. [CrossRef]
45. Graham, J.B.; Stephenson, J.R.; Smith, I.J. Public perceptions of wind energy developments: Case studies from New Zealand. *Energy Policy* **2009**, *37*, 3348–3357. [CrossRef]
46. Ek, K. Public and private attitudes towards "green" electricity: The case of Swedish wind power. *Energy Policy* **2005**, *33*, 1677–1689. [CrossRef]
47. Swofford, J.; Slattery, M. Public attitudes of wind energy in Texas: Local communities in close proximity to wind farms and their effect on decision-making. *Energy Policy* **2010**, *38*, 2508–2519. [CrossRef]
48. Tian, W.; Wang, Y.; Ren, J.; Zhu, L. Effect of urban climate on building integrated photovoltaics performance. *Energy Convers. Manag.* **2007**, *48*, 1–8. [CrossRef]
49. Devine-Wright, P. Beyond NIMBYism: Towards an integrated framework for understanding public perceptions of wind energy. *Wind Energy* **2005**, *8*, 125–139. [CrossRef]
50. Frederiks, E.R.; Stenner, K.; Hobman, E.V. Household energy use: Applying behavioural economics to understand consumer decision-making and behaviour. *Renew. Sustain. Energy Rev.* **2015**, *41*, 1385–1394. [CrossRef]

51. Ertör-Akyazı, P.; Adaman, F.; Özkaynak, B.; Zenginobuz, Ü. Citizens' preferences on nuclear and renewable energy sources: Evidence from Turkey. *Energy Policy* **2012**, *47*, 309–320. [CrossRef]
52. Warren, C.R.; McFadyen, M. Does community ownership affect public attitudes to wind energy? A case study from south-west Scotland. *Land Use Policy* **2010**, *27*, 204–213. [CrossRef]
53. Liu, W.; Wang, C.; Mol, A.P.J. Rural public acceptance of renewable energy deployment: The case of Shandong in China. *Appl. Energy* **2013**, *102*, 1187–1196. [CrossRef]
54. Zografakis, N.; Sifaki, E.; Pagalou, M.; Nikitaki, G.; Psarakis, V.; Tsagarakis, K.P. Assessment of public acceptance and willingness to pay for renewable energy sources in Crete. *Renew. Sustain. Energy Rev.* **2010**, *14*, 1088–1095. [CrossRef]
55. Tsagarakis, K.P.; Bounialetou, F.; Gillas, K.; Profylienou, M.; Pollaki, A.; Zografakis, N. Tourists' attitudes for selecting accommodation with investments in renewable energy and energy saving systems. *Renew. Sustain. Energy Rev.* **2011**, *15*, 1335–1342. [CrossRef]
56. Braito, M.; Flint, C.; Muhar, A.; Penker, M.; Vogel, S. Individual and collective socio-psychological patterns of photovoltaic investment under diverging policy regimes of Austria and Italy. *Energy Policy* **2017**, *109*, 141–153. [CrossRef]
57. Zhai, P.; Williams, E.D. Analyzing consumer acceptance of photovoltaics (PV) using fuzzy logic model. *Renew. Energy* **2012**, *41*, 350–357. [CrossRef]
58. Firestone, J.; Bidwell, D.; Gardner, M.; Knapp, L. Wind in the sails or choppy seas?: People-place relations, aesthetics and public support for the United States' first offshore wind project. *Energy Res. Soc. Sci.* **2018**, *40*, 232–243. [CrossRef]
59. Hui, I.; Cain, B.E.; Dabiri, J.O. Public receptiveness of vertical axis wind turbines. *Energy Policy* **2018**, *112*, 258–271. [CrossRef]
60. Scherhaufer, P.; Höltinger, S.; Salak, B.; Schauppenlehner, T.; Schmidt, J. Patterns of acceptance and non-acceptance within energy landscapes: A case study on wind energy expansion in Austria. *Energy Policy* **2017**. [CrossRef]
61. Gadenne, D.; Sharma, B.; Kerr, D.; Smith, T. The influence of consumers' environmental beliefs and attitudes on energy saving behaviours. *Energy Policy* **2011**, *39*, 7684–7694. [CrossRef]
62. Norton, B.; Eames, P.C.; Mallick, T.K.; Huang, M.J.; McCormack, S.J.; Mondol, J.D.; Yohanis, Y.G. Enhancing the performance of building integrated photovoltaics. *Sol. Energy* **2011**, *85*, 1629–1664. [CrossRef]
63. Wilcox, J.B.; Rossi, P.H.; Wright, J.D.; Anderson, A.B. Handbook of Survey Research. *J. Mark. Res.* **1985**, *22*, 100. [CrossRef]
64. Freund, R.J.; Wilson, W.J.; Mohr, D.L. Data and Statistics. In *Statistical Methods*; Academic Press: Cambridge, UK, 2012.
65. Ho, R. *Handbook of Univariate and Multivariate Data Analysis with IBM SPSS*; Chapman & Hall/CRC: Boca Raton, FL, USA, 2006; ISBN 9781439890226.
66. Schuenemeyer, J.H.; Murtagh, F.; Heck, A. Multivariate Data Analysis. *Technometrics* **2006**, *49*, 103–104. [CrossRef]
67. Jolliffe, I.T. *Principal Component Analysis*, 2nd ed.; Springer: New York, NY, USA, 2002; ISBN 0-387-95442-2.
68. Maxwell, A.E.; Harman, H.H. Modern Factor Analysis. *J. R. Stat. Soc. Ser. A* **1968**, *131*, 615–616. [CrossRef]
69. Ntanos, S.; Kyriakopoulos, G.; Chalikias, M.; Arabatzis, G.; Skordoulis, M.; Galatsidas, S.; Drosos, D. A social assessment of the usage of Renewable Energy Sources and its contribution to life quality: The case of an Attica Urban area in Greece. *Sustainability* **2018**, *10*, 1414. [CrossRef]
70. Kowalska-Pyzalska, A. What makes consumers adopt to innovative energy services in the energy market? A review of incentives and barriers. *Renew. Sustain. Energy Rev.* **2018**, *82*, 3570–3581. [CrossRef]
71. Liarakou, G.; Gavrilakis, C.; Flouri, E. Secondary school teachers' knowledge and attitudes towards renewable energy sources. *J. Sci. Educ. Technol.* **2009**, *18*, 120–129. [CrossRef]
72. Manolas, E.I. Environmental education programs in greek secondary schools: Practices, Problems and promises. In *Environmental Education in Context: An International Perspective on the Development of Environmental Education*; Taylor, N., Littledyke, M., Eames, C., Coll, R., Eds.; SENSE Publishers: Roterdam, The Netherlands; Boston, MA, USA; Taipei, Taiwan, 2009; pp. 83–98.

73. Liefländer, A.K.; Bogner, F.X.; Kibbe, A.; Kaiser, F.G. Evaluating Environmental Knowledge Dimension Convergence to Assess Educational Programme Effectiveness. *Int. J. Sci. Educ.* **2015**, *37*, 684–702. [CrossRef]
74. Mirza, U.K.; Harijan, K.; Majeed, T. Status and need of energy education: The case of Pakistan. In *Energy, Environment and Sustainable Development*; Uqaili, M.A., Harijan, K., Eds.; Springer Werlag: Wien, Austria, 2012; pp. 39–49. ISBN 9783709101094.

© 2019 by the authors. Licensee MDPI, Basel, Switzerland. This article is an open access article distributed under the terms and conditions of the Creative Commons Attribution (CC BY) license (http://creativecommons.org/licenses/by/4.0/).

*Article*

# Evaluation of Co-Existence Options of Marine Renewable Energy Projects in Japan

**A.H.T. Shyam Kularathna [1],\*, Sayaka Suda [2], Ken Takagi [3] and Shigeru Tabeta [2]**

1. Graduate Program in Sustainability Science-Global Leadership Initiative, Graduate School of Frontier Sciences, The University of Tokyo, Room 334, Building of environmental studies, 5-1-5 Kashiwanoha, Kashiwa, Chiba 277-8563, Japan
2. Department of Environment Systems, Graduate School of Frontier Sciences, The University of Tokyo, 5-1-5 Kashiwanoha, Kashiwa, Chiba 277-8563, Japan; tfs.suda.m31@gmail.com (S.S.); tabeta@k.u-tokyo.ac.jp (S.T.)
3. Department of Ocean Technology, Policy, and Environment, Graduate School of Frontier Sciences, The University of Tokyo, 5-1-5 Kashiwanoha, Kashiwa, Chiba 277-8563, Japan; takagi@edu.k.u-tokyo.ac.jp
* Correspondence: shyamkularathna@gmail.com or shyamkularathna@s.k.u-tokyo.ac.jp

Received: 26 April 2019; Accepted: 13 May 2019; Published: 18 May 2019

**Abstract:** Consensus building among local stakeholders is vital for the success of the proposed initial commercial marine renewable energy (MRE) projects in Japan. Even though the literature on stakeholder acceptance highlights the importance of creating local benefits and co-creation options, very few studies and almost no empirical data have been published on the application of non-monetary benefit creation schemes in the context of MRE. Hence, the purpose of this study was to systematically evaluate the possible co-existence options available for Japan's MRE projects through data collected from interviews and questionnaire surveys in two development sites in Nagasaki and Kitakyushu in Southern Japan. To overcome the limitations of data unavailability and uncertainty, the Dempster Shafer Analytic Hierarchy Process (DS-AHP) was used for evaluating the best co-existence strategy out of five potential options. The results indicate that local fisheries prefer the oceanographic information sharing option whereas most of the other stakeholders prefer using local resources to construct and operate the power plant, creating business involvement opportunities for the local community. Analysis of stakeholders' decision behaviors suggests that perceived impacts, knowledge, and values influence the preference decision. In addition to the validation of stakeholder preference of the previously proposed co-existence options with empirical data, this study provides a robust method to further evaluate the potential options with the availability of new data.

**Keywords:** marine renewable energy; co-existence; co-location; Dempster Shafer Analytic Hierarchy Process; multi-criteria analysis

---

## 1. Introduction

Marine renewable energy (MRE) is often considered to be the renewable energy resources that can be extracted from nearshore and offshore areas such as waves, tidal and ocean currents, thermal and salinity gradients and, offshore wind [1–3]. The estimated potential of these MRE resources is significant in comparison to the global demands [2,4]. However, most of the MRE technologies are still in the readiness phase, except for the offshore wind energy sector in some European regions [5]. From the experiences of onshore renewable energy developments and initial developments of MRE projects in Europe, it is identified that overcoming related technological and economic challenges is essential [5] but will not be sufficient for sustainable MRE development if project developers fail to achieve consensus among related stakeholders [6]. However, important differences exist in stakeholder engagement with MRE compared to onshore energy infrastructures [7]. Hence, the interactions between power projects and the local community are significantly different among onshore renewable energy projects and MRE projects. Unlike European MRE developments where different MRE technologies have been developing and testing steadily since the early 1980s, Japan's MRE industry development had an early but very slow start from the 1970s to 1990s. Only few pilot projects were completed in the early 2000s in Japan. However, due to the changes caused by the 2011 Fukushima nuclear disaster, development of the MRE sector accelerated. Japan is aiming to initiate the first commercial offshore wind projects within the first decade of its accelerated MRE development phase. However, the required policy regulations, as well as public perceptions, are yet to be implemented and understood for MRE developers to ensure the success of the planned commercial projects.

*1.1. Problem Definition*

Existing regulations are unclear and scattered among many agencies and no clear marine spatial planning approach is used in Japan. Nearshore areas are generally allocated to fishery industry and given to regional fishery unions with fishing rights. Offshore areas are being used on a shared permission basis. Regulations on using marine areas for MRE projects have been recently introduced [8]. The regulations and guidelines applicable to considered marine areas differs significantly according to the nature of the regulating agency. The general guideline states that MRE project developers have to achieve consensus of relevant stakeholders prior to submission of the project development proposal to obtain permission to use marine areas (exclusive use of the marine area for 30 years) for their projects [9].

Local project impacts and local benefits are one of the basic criteria being considered in the process of local acceptance decision making [10]. Reducing the risk of negative environmental impact has been identified and discussed frequently during the consensus building process. However, a little work has been completed on potential co-existence strategies that can be used to create local benefits from the introduction of MRE projects, even though providing community benefits can increase levels of local support through improving individual perceptions of MRE projects [11]. Previous studies have shown that community benefits are unlikely to increase local stakeholder support when bribery perceptions are salient [12]. Further institutionalizing community benefit schemes has the potential to reduce bribery perceptions [13]. Empirical results from a potential offshore wind farm in the United Kingdom suggested that local stakeholder support is greater if the community benefits result from an institutionalized policy guidance in comparison to the community benefit schemes created as a voluntary act by the project developers [14]. The basic problem of creating local community benefit schemes is that it normally refers only to additional voluntary measures provided by the developers, which leads to additional costs to already expensive MRE projects. Most monetary benefit creation strategies lead to higher costs to the developer, which is directly proportional to the number of beneficiaries [15]. To be acceptable by the project developers, the proposed solutions must not follow the same path of monetary benefit schemes. Conversely, it would be ideal if the additional local benefits can be created with the help of the infrastructure developed for the MRE projects (i.e., use the co-benefits of MRE infrastructure). However, there are no previous examples in the Japanese MRE context and limited literature even from other contexts are available for project developers to evaluate such potential benefit creation and co-existence strategies. Developers have to manage a variety of stakeholder groups and are generally not equipped to balance all their requirements. Hence, it is of utmost importance to evaluate options to create local project benefits and a win-win situation among all the local stakeholders of MRE projects.

Non-monetary benefit schemes have been identified as options to create community benefits from the learnings of onshore renewable energy projects [6]. Local fishery industry is the most likely to be directly impacted by MRE project deployments [16] and project structures are usually built away from community; the related social conditions as well as potential benefit creation strategies of MRE projects tend to deviate from its onshore counterpart. The Research Institute of Ocean Economics (RIOE) in Japan has proposed some options to create benefits for local fisheries from offshore wind projects [17]. Almost no literature or practical examples of application of non-monetary local benefit schemes are available in the context of MRE development in coastal communities. The second aspect of this problem is the lack of knowledge and experience as local MRE sector is still in the technology readiness phase. Hence, consensus building and co-existence strategy selection process have to be completed with a significant level of uncertainty. The overarching problem focused in this study is: what is the best local benefit creation and co-existing strategy preferred by stakeholders to create a win-win situation for the introduction of commercial level MRE projects in Japan?

*1.2. Case Study Sites*

This study is based on a data set collected from two case study sites from Nagasaki and Kitakyushu in Southern Japan (Figure 1) that represents the best examples of private companies trying to initiate commercial-level MRE projects after experiencing success with government funded demonstration pilot projects. The Ministry of Environment in Japan has been conducting pilot MRE projects in Nagasaki since 2010 (and developed into Japan's first full-scale grid-connected 2 MW floating offshore wind turbine) near the Goto Islands, about 100 km off the main Nagasaki city [18]. Initially, the project owners received some local concern about the development of MRE devices and testing in real sea conditions due to the perceived negative environmental impacts. Given the results of the pilot projects, local stakeholder acceptance of Nagasaki MRE development has increased significantly. A non-profit organization (NPO), Nagasaki Marine Industry Cluster Promotion Association (NaMICPA), comprised of more than 50 public and private entities related to marine industries, was established in 2014 with the aim of supporting the development of marine industries including the MRE sector [19]. A proposal to build the first commercial MRE project, a 22 MW offshore wind farm has been submitted by a private company with expected commencement in 2019 [20]. The Naru strait, which lies between two smaller islands, has been identified as a potential site for tidal energy projects and authorities are in the process of establishing a marine energy test center similar to the European Marine Energy Centre (EMEC) in Scotland [21].

Similar to the inception of Nagasaki MRE projects, government agencies have started testing the feasibility of offshore wind energy development in the Hibiki Sea area in Kitakyushu, Japan in 2012 [22]. A consortium comprised of local industries, the Hibiki Wind Energy Group [23], won the bid to build the first large-scale offshore wind farm in Kitakyushu in 2017 [20], which is planned to start in 2022.

In the case of tidal energy projects amongst the Goto Islands in the Nagasaki case study area, the approval of the local fishery association is necessary since the tidal strait was already declared as a marine area with fishing rights. However, the offshore wind project in the Nagasaki case study is planned in the general offshore area where any marine user can use the area on a shared permission basis with the approval of local authorities. In contrast, the Kitakyushu offshore project is planned in marine areas governed by port law; hence, the local port authority has the exclusive control over the development site. Despite the differences in the required legal requirements, both Nagasaki and Kitakyushu MRE developers are compelled to search for means to improve public acceptance of their proposed commercial MRE projects.

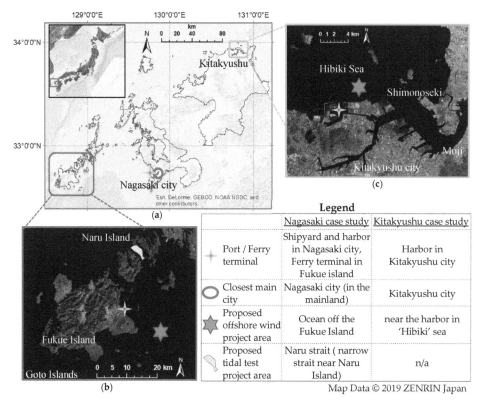

**Figure 1.** Location of Nagasaki and Kitakyushu MRE development sites in Southern Japan: (**a**) Location of two case studies; (**b**) Fukue and Naru Islands (in Goto Islands) in Nagasaki case study; (**c**) Kitakyushu city, Moji, and Shimonoseki area in Kitakyushu case study.

## 2. Materials and Methods

*2.1. Data Collection*

The main data were collected using key stakeholder interviews and a questionnaire survey in the main communities near the Nagasaki and Kitakyushu MRE project sites. We conducted 20 key stakeholder interviews with the local fishery union representatives (7), project developers (3), local government agency respondents (3), NaMICPA non-profit organization representatives (4), and environmental observation teams including related researchers (3). Key stakeholder interviews were conducted in a semi-structured format focusing on the potential benefit creation options found in the literature.

The questionnaire survey was conducted in coastal communities of Fukue Island, Naru Island, and Nagasaki city area, representing the Nagasaki case study, whereas the Kitakyushu city area, Moji, and the Shimonoseki area represented the Kitakyushu case study. A total of 77 responses were selected as complete and valid for further analysis (Table 1). The questionnaire survey included questions to elucidate the demographic information of the respondents, potential co-existence options and evaluation criteria, pair-wise comparison of evaluation criteria, and preference of identified co-existence strategy based on each criteria.

**Table 1.** Summary of valid questionnaire survey respondents.

| Stakeholder Group | Number of Respondents | | |
|---|---|---|---|
| | Total | Nagasaki | Kitakyushu |
| Local fishery [1] | 15 | 9 | 6 |
| Developers/Construction sector | 7 | 6 | 1 |
| Civil servants | 14 | 14 | - |
| Tourism and shipping industry | 2 | 1 | 1 |
| Health and welfare | 4 | 3 | 1 |
| Non-profit organizations (NPO), Service sector and others | 19 | 14 | 5 |
| Not indicated | 16 | 13 | 3 |
| | 77 | 60 | 17 |

[1] Fishery union officials and general fishers in the area.

## 2.2. Data Analysis

Overarching data for MRE co-existence option evaluation was analyzed using the Dempster Shafer Analytic Hierarchy Process (DS-AHP) multi-criteria decision making model. The entire process can be summarized into three main steps: (1) identification of potential co-existence options, (2) multi-criteria analysis (MCA) of identified co-existence options and, (3) evaluation of the stakeholder preferences using the DS-AHP model based on the MCA results.

### 2.2.1. DS-AHP Multi-Criteria Decision Making Model

The main data analysis method involves two fundamental decision making techniques: Analytic Hierarchy Process (AHP) and Dempster Shafer Theory (DST). AHP [24,25] is one of the most widely used multi-criteria decision making approaches in many disciplines. The main limitation of AHP is its requirement for pair-wise comparison for each option pair combination, which makes the process impractical when many options must be considered and there is a significant level of data unavailability or uncertainty. DST, which is based on the belief function, is used to overcome the limitation of handling uncertainty in the standard AHP method. DST originated from the methods developed by Dempster to estimate upper and lower probabilities [26,27] and improvements were added by Shafer [28]. DST is widely applied in the field of machine learning and artificial intelligence. The combination of AHP and DST as a multi-criteria decision making model is also known as the Dempster Shafer Analytical Hierarchy Process (DS-AHP). DS-AHP is a more robust framework suitable for decision making under uncertainty where AHP is used for rating decision criteria and DST is used for evaluation of decision options using the weighted criteria [29–31]. Only the basics of DS-AHP are explained here. Please refer the original literature [29–31] for more information.

### 2.2.2. Interpretation of DS-AHP Calculation

Let $\Theta = \{h_1, h_2 \ldots h_n\}$ be a collectively exhaustive and mutually exclusive finite set of $n$ hypotheses or propositions, which is also called the frame of discernment. The basic probability assignment (bpa), is a function $m: 2^\Theta \rightarrow [0,1]$ that also satisfies the requirement $m(\phi) = 0$ and $\sum_{A \subseteq \Theta} m(A) = 1$, where $\phi$ represents the empty set and $2^\Theta$ represents the power set of $\Theta$. The assigned probability of any sub set $y$ of frame of discernment $\Theta$, (i.e., $y \subseteq \Theta$) is denoted by $m(y)$. $m(y)$ represents the exact belief in the proposition depicted by $y$. The assigned probability for the frame of discernment $\Theta$ (i.e., $m(\Theta)$), represents the global ignorance within the bpa [30].

In the DS-AHP model, $m_i(y)$, i.e., the bpa value for decision alternative(s) $y$ with respect to the decision criteria $i$, is calculated using Equation (1):

$$m_i(y) = \frac{a_y W_i}{\sum_{j=1}^{d} a_j W_i + \sqrt{d}} \quad \text{and} \quad m_i(\Theta) = \frac{\sqrt{d}}{\sum_{j=1}^{d} a_i W_i + \sqrt{d}} \qquad (1)$$

where $a_y$ denotes the user preference value (1–7 preference scale from lowest preference 1 to highest preference 7), $W_i$ represents the weight assigned to the considered decision criteria by pair-wise comparison using the standard AHP method [25] and $d$ represents the number of decision alternatives judged by the decision maker [31].

Basic probability assignments are considered as evidence and can be combined using Dempster's rule of combination, provided that information sources are independent. Criteria-wise preference probabilities can be combined and decision maker-wise preferences can be calculated using Dempster's rule of combination in Equation (2):

$$m_{i \oplus j}(y) = \begin{cases} 0 & ; y = \phi \\ \frac{\sum_{A_p \cap A_q = y} m_i(A_p) \, m_j(A_q)}{1 - \sum_{A_p \cap A_q = \phi} m_i(A_p) \, m_j(A_q)} & ; y \neq \phi \end{cases} \quad (2)$$

where $m_{i \oplus j}(y)$ denotes the combined preference probability with respect to decision criteria $i$ and $j$. This combination rule is used again to aggregate the individual decision maker's preference levels to derive the group preference, taking each decision maker as a criteria [31].

Belief level, denoted by $Bel(y)$ represents the confidence or exact support for the proposition $y$ or the confidence level that hypothesis $y$ is true. Plausibility level, denoted by $Pls(y)$, represents the possibility of support for proposition $y$ or the maximum amount of confidence that could be placed on $y$. Both belief and plausibility are functions: $2^\Theta \rightarrow [0,1]$ and constitute the interval of support for the considered proposition $y$. The two functions are related to each other by $Pls(y) = 1 - Bel(\overline{y})$ where $\overline{y}$ represents the complement of $y$. The interval between belief and plausibility levels represents the uncertainty level because $[Bel(y), Pls(y)]$ represents the lower and upper bounds of the probability by which the considered proposition $y$ is supported [28,32]. The final belief level and the plausibility levels are calculated by Equation (3).

$$Bel(S) = \sum_{B \subseteq S} m(B) \; \forall S \subseteq \Theta \quad \text{and} \quad Pls(S) = \sum_{B \cap S \neq \emptyset} m(B) \; \forall S \subseteq \Theta \quad (3)$$

### 2.3. Option Identification and DS-AHP Decision Hierarchy

The potential co-existence options were basically identified by analyzing the proposal made by the Research Institute of Ocean Economics in Japan [17], and systematically understanding the real project stakeholders' perceptions of each proposal via key stakeholder interviews and other related literature. After identifying the potential and the applicability, the proposed co-existence options were categorized into five main options: (O1) sharing in-situ, real time oceanographic information; (O2) using MRE structures as artificial reefs and support structures for commercial fishing; (O3) co-location with other industries such as leisure and tourism, aquaculture, etc.; (O4) sharing generated electricity for local users at a subsidized rate; and (O5) use of local resources to construct and operate the power plant, creating business involvement opportunities. Table 2 provides a summary of the option identification. In the next step, local relevance, expected impacts, and limitations of each identified option were analyzed with the data from key stakeholder interviews.

From the key stakeholder interviews, we identified that most of the considered co-existence options have not been used similarly, even in other contexts. Hence, the potential interactions and impacts were still unknown, making direct quantitative evaluation unreliable. Instead, the preference was evaluated based on perceived potential impacts identified during key stakeholder interviews. A broader set of criteria, i.e., economic impacts, environmental impacts, and stakeholder engagement and other social impacts, was selected as the preference decision criteria in the decision hierarchy shown in Figure 2a. The individual stakeholder-wise decisions to aggregate to a group decision were combined using the equally weighted decision makers approach, where individual decision makers of the considered group are considered as equally weighted decision criteria according to the DS-AHP group decision methodology [31] (Figure 2b).

Table 2. Summary of co-existence strategy identification.

| Selected Co-Existence Options | Options Proposed by Research Institute of Ocean Economics (RIOE), Japan [17] | Key Findings from Other Literature |
|---|---|---|
| O1. Providing real-time, in-situ ocean information from MRE farms | Providing marine information in real-time | Japanese marine users receive ocean information from satellite observations, buoys, and other user specific monitoring platforms [17,33]. Stakeholders have different oceanographic information demands [34]. Direct economic valuation and cost benefit analysis of ocean information is impractical [35]. Marine energy is mostly harvested in murky and high energetic places where conventional data acquisition techniques are impractical [36]. |
| O2. Using MRE structures as artificial reefs and support structures for commercial fishing | Use MRE structures as artificial reefs for Nurseries/Fishing | Constraints, opportunities, and perceptions of co-locating offshore wind farms and fisheries [37,38], mitigation agenda for fishing effort displacement [39]. |
| | Using MRE structures to support fishery gears | Potential for co-location of passive gear fisheries with offshore wind [40]. Potential for and limitations of co-locating fisheries inside offshore wind farms [41]. |
| O3. Co-location with other industries such as leisure, tourism, and aquaculture | Co-location with aquaculture facilities (e.g., Fish, Oyster, and Algae) | Co-locating offshore wind farms and aquaculture facilities [38,42–45]. Device placement has many other technical requirements [46] |
| | Co-location with leisure facilities (e.g., diving, recreational fishing etc.) | Potential for limited entry recreational fishery in wind farms [47], snorkeling, tourism [48], angling, and yachting [43,49] |
| O4. Sharing generated electricity for local users at a subsidized rate | Use of electricity generated to power fishery port facilities and electric boats | Proposal for using wind energy to power fishery ports [50], harbors [51], desalination plants [52]. |
| O5. Use of local resources to construct and operate the power plant creating business involvement opportunities | Project participation by using fishery boats for construction and maintenance of the power plant | Use of fishing vessels for offshore energy projects [53]. Availability of crew and vessels is an important factor influencing the planning and cost of maintenance of MRE. Laws and regulations also influence MRE operation and maintenance (O&M) [54,55]. |
| | Project participation by providing investment opportunities in MRE business | Creating business investment opportunities as an acceptance improvement measure [56]. Local ownership or financial participation contribute to the acceptance of MRE projects [57]. |

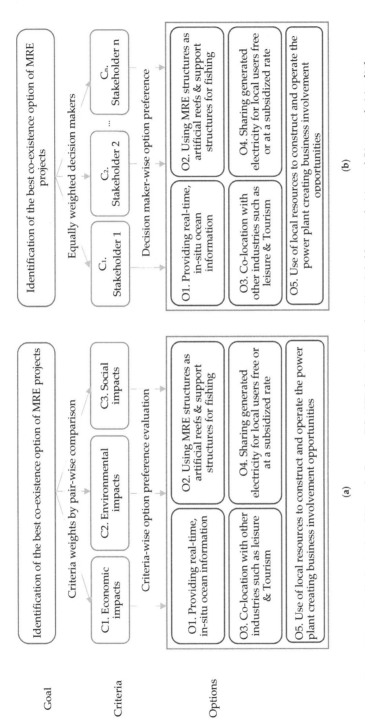

Figure 2. Dempster Shafer Analytical Hierarchy Process (DS-AHP) decision hierarchy: (a) stakeholder-wise decision and (b) group decision [31].

## 3. Results

As the last part of option identification, non-monetary co-existence options identified from literature were confirmed via stakeholder interviews. Stakeholder interviews were conducted in a semi-structured format that focused on the expected potential impacts, related costs and benefits, and the related risks and limitations of the proposed co-existence options. Table 3 summarizes the main points provided by the key stakeholders during the interviews.

Table 3. Summary of stakeholder interviews on expected impacts, related costs and benefits, and risks and limitations of proposed co-existence options.

| Co-Existence Option | Expected Impacts, Related Costs and Benefits, Risks and Limitations |
| --- | --- |
| O1. Providing real-time, in-situ ocean information from MRE farms | <ul><li>Real-time in-situ ocean information is valuable to the marine users due to travel cost reductions, risk reductions, and improvements in commercial marine industries such as fisheries (by efficient fishing ground selection, stock estimations, etc.) and navigation (improvements in safety, route planning, etc.)</li><li>Can be identified as a co-benefit of the MRE projects since most commonly-required ocean information can be generated from the Condition Monitoring System (CMS) of the power farm.</li><li>Stakeholder engagement can be improved since many stakeholders directly or indirectly use ocean information.</li><li>The additional cost to developers is insignificant (if there is no ocean monitoring equipment to be installed in addition to the power plant's standard CMS) and not proportional to the number of beneficiaries due to the existence of cheap information dissemination methods.</li><li>Equality and scalability can be improved if the governance of information sharing is well-maintained.</li><li>There is a risk of stakeholder conflicts due to the exposure of marine information that is considered trade secret (such as fishing grounds). Information about the marine environment can lead to better eco-system management as well as unsustainable exploitation of marine resources (such as over fishing) unless there is proper governance of shared ocean information.</li></ul> |
| O2. Using MRE structures as artificial reefs and support structures for fishing | <ul><li>Artificial reef effect and resulting positive spillover effect to the surrounding fishing grounds can be considered a co-benefit of MRE projects.</li><li>Use of sub-structures to support fisheries can be a benefit if there is no significant additional cost to the developer and fishing gear does not adversely interact with the MRE devices.</li><li>Only certain types of fishers can benefit since many fishing methods are being used in the case study areas.</li><li>Scalability is directly dependent of the size of the MRE farm.</li><li>There is a high possibility of increasing the initial construction costs as well as O&M costs if MRE structures are used as support structures for fishing operations. Impact to the overall Levelized Cost of Energy (LCOE) and the net benefits to the fishery industry should be considered when conducting a detailed cost-benefit analysis for this option.</li><li>Artificial reef effects caused by bio-fouling as well as fishing operations near MRE devices can pose significant operational risks and unforeseen problems.</li></ul> |

Table 3. *Cont.*

| Co-Existence Option | Expected Impacts, Related Costs and Benefits, Risks and Limitations |
|---|---|
| O3. Co-location with other industries such as leisure, tourism, and aquaculture | <ul><li>Aquaculture is one of the best co-location options; however, it depends on how fishing gears can be used with MRE structures. With the combination of reef effects and remote monitoring facilities (e.g., detection of fish within MRE farms [36]), aquaculture facilities combined with MRE farms seems to be an attractive solution.</li><li>Local tourism can be improved by having visible MRE projects as well as organizing boat excursions to the power farm areas. Reef effect creates an environment conducive for snorkeling and diving.</li><li>There should be a practical method of regulating the interactions to maintain the safety and efficiency of both industries.</li><li>Due to the nature of operations, such as travel planning, aquaculture facilities (specially seaweeds culture) and leisure facilities have the same characteristics that differ from typical large-scale fishing.</li><li>Operations performed in marine environments near MRE farms can pose significant risks to the MRE devices as well as the involved personnel.</li><li>LCOE can be impacted by additional construction or O&M costs due to co-location attempts.</li></ul> |
| O4. Sharing generated electricity for local users at a subsidized rate | <ul><li>Local fishery harbors and fish processing plants can be the best candidates for receiving subsidized electricity.</li><li>Under current regulations, it is illegal for the utility company to differentiate the electricity rates based on other factors. Hence, limiting the number of beneficiaries is difficult unless clear policy-level guidance is introduced.</li><li>Additional costs are directly proportional to the number of beneficiaries, thus limiting scalability and economic viability.</li><li>Offshore charging points for electric boats (like charging stations for electric vehicles on land) can be created in the future; however, those technologies are too uncertain and impractical given existing costs.</li><li>LCOE can be impacted by additional construction or O&M costs due to potential additional requirements of local electricity grid management.</li></ul> |
| O5. Use of local resources to construct and operate the power plant creating business involvement opportunities | <ul><li>Shipping vessels can be used for logistic purposes during the environmental impact assessment, construction, as well as maintenance phase of the power farm. Local fishers can be recruited for monitoring purposes in the offshore area.</li><li>If the local fishery union can invest in the project, the sense of ownership can lead to a better performance of fishers as guards of the power farm.</li><li>However, the local capacity within fisheries is limited and legal regulations have to be adopted accordingly.</li><li>Local ports, the steel industry in Kitakyushu, and the ship building industry in Nagasaki can be strategic partners of future MRE projects.</li><li>LCOE can be positively impacted by using local resources from already established sources and industries. However, LCOE can be adversely impacted if the initial MRE projects have to invest in capacity building of the local sources to make them qualified and competent enough to be involved with the MRE projects.</li></ul> |

## 3.1. Qualitative Multi-Criteria Analysis

At the end of the interviews, stakeholders' perceptions about the proposed co-existence options were analyzed using multi-criteria analysis (MCA). Since no common unit of measurement exists for each criterion selected, and the stakeholders did not have a quantifiable amount for each criterion, multi-criteria evaluation was qualitatively completed based on their perceptions and reported perceived impacts.

Economic aspects were considered using three sub-criteria. The project co-benefit criterion measures the extent of the considered co-existence option being a co-benefit of the MRE project. Co-benefit was roughly contextualized as all secondary benefits of the MRE project other than the intended benefit of sustainable renewable energy supply. The second sub-criterion under economic aspects was the measure of variable cost to the developer, i.e., the amount of additional costs the developer has to incur for each additional beneficiary. The lower the variable cost, the lower the project cost. Since the sea area has vague ownership due to the lack of a well-established marine spatial plan, limiting the number of beneficiaries is practically difficult. This is the main reason for the unviability of monetary compensation schemes. The third economic sub-criterion is related to scalability of the solution without adding significant developer costs. Indirectly, it can be described as the ability to provide the same level of service without adding significant fixed costs to the developer. Impacts to marine environment and greenhouse gas (GHG) emission levels were the main ideas highlighted during the interviews regarding environmental impacts. Social implications were measured by three common social criteria: stakeholder engagement, level of incentives to the stakeholder, and equality. The level of incentives can be an indirect and qualitative measure of the perceived benefit levels. Equality is considered between all the stakeholder groups in the local context. Table 4 summarizes the qualitative MCA of the selected co-existence options. Qualitative MCA results that had no common unit of measurement were converted to three quantitative measures indicated by ✓ (affirmative/positive impacts), - (not sure), and x (non-affirmative/negative impacts). The number of repetitions (up to three times) of the symbols ✓ and x represents the degree of agreement (tendency to somewhat agree, agree, and strongly agree, respectively) for all the stakeholder interviews considered cumulatively. This level assignment of was completed based on the authors' best estimates and based on the characteristics of the interview results such as the frequency of mentioning the considered point and the level of confidence of the interviewee regarding the considered point.

The main limitations of the considered co-existence options identified during key stakeholder interviews are shown in the last row of Table 4. For example, the main concerns mentioned regarding the ocean information sharing option were: how the shared information will be used in the context of competitive fishing ground selection, who will be given the information because some fishery groups maintain knowledge about fishing grounds as a local trade secret and fishers from outside areas also have the possibility to use the same fishing area, and if the new information will cause sustainable stock management or over exploitation of fishery resources. All these concerns have to be handled by establishing good governance for using the shared information. Only a certain type of fishers can benefit from the second option of using MRE structures as artificial reefs or support structures for fishing gear. Hence, unequal cost-benefit distribution and limitations of scaling the benefits to other stakeholders were mentioned as limitations of the second option. Since there are no prior examples of combining aquaculture or leisure facilities with other offshore activity, there is a significant uncertainty for the feasibility of the third option, even though the possibility was recognized by the stakeholders. Local utility company representatives indicated that they are legally bound to maintain equality in terms of pricing the electricity for their customers, so the electricity rate for different customers or stakeholder groups cannot be significantly differentiated. Fishery union representatives and the developers identified the limitations of the fifth option as the requirement of specialized skills and other resources to become involved with the MRE sector. For example, even though the fishery vessels can be used as power plant monitoring resources (at a certain distance), they might not be capable of being used as a logistic means to reach or repair the MRE devices. The limitations of local capacity were identified as the main limitation of the fifth co-existence option of using local resources to construct, maintain, and operate the power plant and creating business involvement opportunities.

**Table 4.** Multi-criteria analysis (MCA) of co-existence options.

| Key Criteria | | O1. Providing Real-Time, In-Situ Oceanographic Information from MRE Farms | O2. Using MRE Structures as Artificial Reefs and as Fishery Support Structures | O3. Co-Location with Industries like Leisure, Tourism, and Aquaculture. | O4. Sharing Generated Electricity for Local Users at a Subsidized Rate | O5. Use of Local Resources to Create Business Involvement Opportunities |
|---|---|---|---|---|---|---|
| C1. Economic impacts | Project Co-benefits | ✓✓✓ | ✓✓ | ✓✓✓ | x | ✓✓✓ |
| C1. Economic impacts | Cost not proportional to the No. of beneficiaries | ✓✓✓ | x | ✓ | xx | ✓ |
| C1. Economic impacts | Scalability | ✓✓✓ | - | - | x | ✓ |
| C2. Environmental impacts | Marine environment | - | - | ✓ | - | ✓ |
| C2. Environmental impacts | Emissions | ✓ | ✓✓ | - | ✓✓ | ✓ |
| C3. Social impacts | Stakeholder engagement | ✓✓ | ✓✓ | ✓✓ | ✓✓✓ | ✓✓✓ |
| C3. Social impacts | Stakeholder incentives | ✓✓ | ✓ | ✓ | ✓✓✓ | ✓✓ |
| C3. Social impacts | Equality | ✓✓ | - | ✓ | - | ✓ |
| | Main limitations | Lack of information sharing governance | Limited scalability and unequal cost benefit distribution | Uncertainty on economic feasibility with the adjustments required | Legal barriers and limiting number of beneficiaries | Limited local capacity |

Note: ✓: Affirmative/positive impacts, -: not sure, x: non-affirmative/negative impacts (Ratings were assigned according to the cumulative stakeholder inputs).

### 3.2. Stakeholder Group-Wise Group Decision

The next step in the co-existence option evaluation involved using the results of MCA with the DS-AHP decision making model according to the selected decision hierarchy (Figure 2) to identify the optimal solution. Figure 3 indicates the criteria weights (obtained by pair-wise comparison as in AHP method) of the selected criteria (in the left column), and the final belief and plausibility levels of support for the considered co-existence options (in the right column) for both case study areas. The responses were grouped considering the prominence assigned to the stakeholder group as well as the unique characteristics of their responses. Respondents representing the local fisheries assigned a higher weight to economic and environmental impacts (C1 and C2, respectively) than the social impacts (C3) (Figure 3a). The fisheries are the main stakeholder group who frequently require oceanographic information for their daily industrial activities. Interviews with fishery unions indicated the value of subsea information for estimating fish stock, fishing ground, viable catch, and the safety of marine activities. All these factors support their preference of considering oceanographic information as the best option (Figure 3b). Even though they were interested in the fifth option, fishers also raised the question about the real potential of being involved with the MRE project developments and operation, because they have a better understanding of what is required to work in offshore conditions based on their experience. Interviews with fishers showed that fishing vessels can be used for logistic purposes during the environmental impact assessment phase and maintenance phase of the power farm. The potential of recruiting local fishers for monitoring purposes of the power plant was also mentioned. However, fishery union leaders identified that the vessels used for local fishing may not be suitable for MRE projects and the ageing fishery community may not be willing to accept new challenges related to MRE projects. However, this type of interaction with fishery and the MRE industry may attract the younger generation to the fishery industry. In addition to the above factors, the low weight assigned to the social impacts contributed to not selecting the fifth option as a preferred option. The second best alternative for fishers was the second option; however, its preference level was significantly lower than that for the first option. Fishers identified that they can benefit from the artificial reef effects, which have a spillover effect on the surrounding fishing grounds. Fishers indicated that they can reduce costs related to their fishing gear setups (such as fixed nets) if they receive structural support from the MRE structures. The value of real-time in-situ ocean information was again highlighted when the fishers discussed their fishing methods and fishing gear, such as the ability of local fishers to protect their fixed net setups, aquaculture setups, etc., in the event of a sudden ocean currents, commonly known as Kyucho in Japan [58,59]. Another advantage of in-situ ocean information is the ability of fishers to predict the ocean conditions and decide if the fishing gear is suitable before travelling to the area. Further analysis of fisheries preference is discussed in the next section due to their importance to the stakeholder group among all other stakeholders as well as their unique decision behavior.

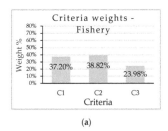
(a)

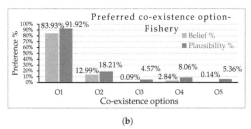
(b)

Figure 3. Cont.

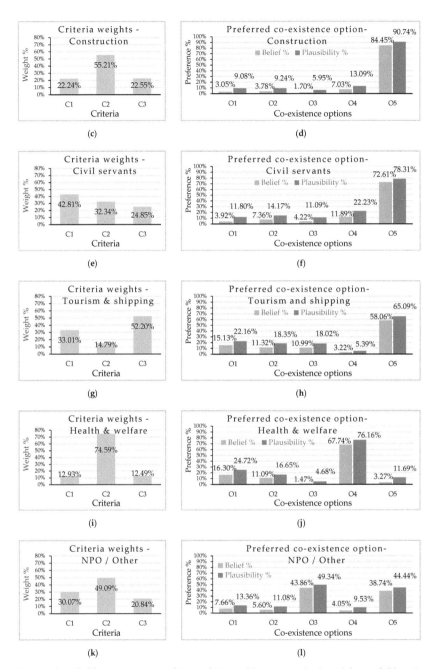

**Figure 3.** Stakeholder group-wise preference decision (**a**) average criteria weights and (**b**) option preference of fisheries; (**c**) average criteria weights and (**d**) option preference of construction sector respondents; (**e**) average criteria weights and (**f**) option preference of civil servants; (**g**) average criteria weights and (**h**) option preference of tourism & shipping industry respondents; (**i**) average criteria weights and (**j**) option preference of health and welfare sector respondents; (**k**) average criteria weights and (**l**) option preference of NPO and other respondents.

Figure 3c,d indicate the criteria weights assigned by and the final preference of the respondents from the construction industry. These respondents are expected to be involved with the MRE projects during its development phase. These results can generally represent the opinion of future MRE project developers. They assigned the highest weight to the environmental impacts. However, the environmental impacts of most of the considered co-existence options are either not known or insignificant (Table 4). The high weight assigned to environmental impacts does not represent the final preference level where the highest preference was for the fifth option. Interviews with the project developers indicated that there is a high possibility of involving local fishers through the local fishery union for the initial stages of MRE project development, such as using their fishing vessels to conduct surveys and environmental impact assessment. The fifth option is the only option that can be directly employed for project development so the developers directly benefit from it. According to the project developers, there can be long term benefits in terms of improving LCOE due to the use of local resources and developing local supply chain industries, even though additional initial investments could be required for building local capacities to meet the requirements of the MRE industry.

Figure 3e,f indicate the criteria weights assigned by and the final preferences of the respondents from the civil service sector. They assigned the highest weight to the economic impacts criterion. They selected the fifth option as the best option among the options. Interviews conducted with local government officers and other civil servants like school teachers indicated that they have no direct involvement with the marine affairs. We separately analyzed the results from the health and welfare sector respondents due to unique characteristics that will be explained later. The group of civil servants considered in Figure 3e,f can be approximated to the inland urban communities that have a vague idea that MRE projects may result in high energy costs and the local community should be given the opportunity to improve their economy.

Respondents involved with the local tourism industry and shipping industry assigned significant weight to the social impacts criterion (Figure 3g), which is comprised of stakeholder engagement, incentives, and equality. The most preferred option was the fifth option: using local resources for MRE project development and creating business involvement opportunities (Figure 3h). The literature as well as key stakeholder interviews indicated the potential for collaborating with these sectors according to both the third and fifth options. However, local respondents had no experience with how MRE projects can collaborate with local tourism industry as indicated by the third option. The high weight assigned to the social impact aspects with the current level of perception might be the reason for their preference for the fifth option over the third option.

The respondents from health and welfare sector had a unique perception of MRE options, even though they can be considered as civil servants in general. This group had in-depth knowledge and experience with human health impacts compared with other civil servants, as indicated by the high weight (74.59%) assigned to the environmental impacts criteria, which was the highest amongst all three criteria weightings for every other stakeholder group (Figure 3i). They can represent the general inland communities given the minimum interaction with marine affairs. Interviews with representatives indicated their concerns about possible low frequency noise and its impact on human health. However, there is no evidence about the impact of low frequency noise from the onshore wind turbines currently installed in their locality. More justifiable reasons for the selection of the fourth option, i.e., sharing generated electricity as the best option as indicated in Figure 3j, would be the expectation that it will reduce the dependency on conventional non-renewable energy sources (like coal), which would reduce GHG emissions and the expectation of reducing the current economic burden caused by the high electricity demand.

Figure 3k,l indicate the criteria weights assigned to and final preference of the respondents from local nonprofit organizations and other community organizations. This group indicated environmental impacts as the most important criterion but selected both the third and fifth options as the preferred options. Since most of these respondents were working closely for the revitalization of the local economy, they expected positive impacts from business involvement opportunities with the new MRE sector. According to the discussions with local hotel owners, they expected to revitalize the local tourism industry via future MRE projects. They indicated that there has been a slight improvement in their businesses due to external people visiting the remote islands because of these project developments.

From the results of the stakeholder group-wise option preferences shown in the Figure 3, no solution clearly meets the preferences of all the stakeholder groups. Stakeholder preferences were significantly related to the expected individual costs and benefits as well as the level of knowledge and interaction with the marine activities. Hence, it was important to further analyze the local preferences according to other factors such as geographical area.

*3.3. Geographical Area-Wise Group Decision*

The area-wise analysis results shown in Figure 4 show that there was no significant preference identification for most of the areas (refer to Figure 1 for the geographical locations of the considered areas). Few area-specific factors were identified related to this area-wise preference decisions, which indicates that preference behavior was more dependent on the stakeholders' occupations than area-specific factors. The main reason for not identifying a clear preference decision in Fukue, Nagasaki, and Shimonoseki areas was that respondents represented number of occupations in these groups. Similarly, most respondents from Kitakyushu were from the fishery industry and most respondents from Moji area were from the health and welfare sector. So, a similar preference pattern can be expected from both area-wise preferences and occupation-wise preferences in respective cases. However, key stakeholder interviews provided some information that supports the decision behaviors in the Naru, Kitakyushu, and Moji areas.

The fishery industry, which is the main traditional industry on Naru Island (near the proposed tidal energy project in Nagasaki case study area), is declining rapidly due to the ageing society and inability to attract the younger generation towards the fishing industry. Interviews with the Naru fishery union representatives also mentioned that there are almost no fishing efforts in the Naru strait due to the high tidal current velocity. Hence, they do not expect to interact much with the tidal energy project. Fishers from Naru Island acknowledged that their fishing efforts could benefit from real-time in-situ ocean information provided according to the first option, by estimating the high tidal current conditions that are unique to their area. Naru fishers identified that they could extend their fishing grounds to the high tidal current areas in the Naru strait if they know the exact conditions of the tidal velocity. Such benefits could be provided even with the second option where MRE structures could help fishing in high velocity tidal streams. However, they do not expect much benefit in terms of fishery due to the diminishing nature of the local fishery industry. Despite most of the Naru respondents being fishers, their preferred strategy was the fifth co-existence option, which was using local resources to construct and operate the power plant, creating business involvement opportunities (Figure 4b).

The Kitakyushu respondents' group preferred the first option of sharing oceanographic information (Figure 4h). This may be because the local harbor, which is a powerful stakeholder in the area, values ocean information more than the other non-fishery stakeholders. The respondents from the Moji area (in the Kitakyushu case study area but away from the MRE project area) preferred the fourth option of sharing generated electricity at a subsidized rate (Figure 4j). Their preference decision can be supported by the fact that the electricity supply could being the only direct impact of the MRE projects, which is relatively far from their dwellings.

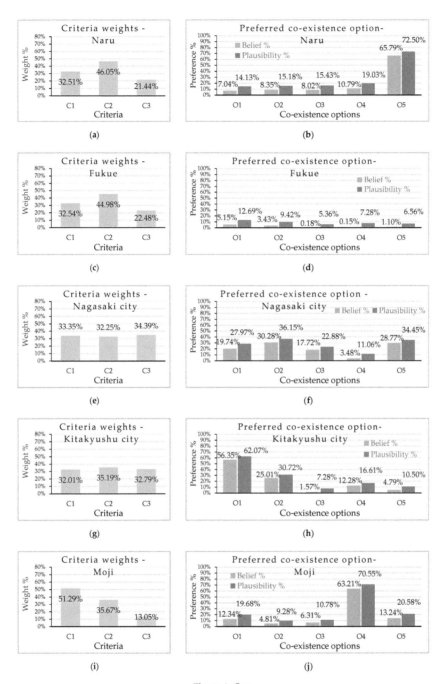

Figure 4. Cont.

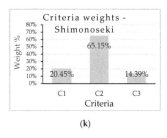 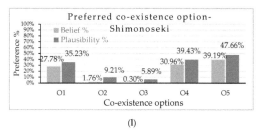

(k) (l)

**Figure 4.** Case study area-wise preference decision (**a**) average criteria weights and (**b**) option preference of respondents from Naru; (**c**) average criteria weights and (**d**) option preference of respondents from Fukue; (**e**) average criteria weights and (**f**) option preference of respondents from Nagasaki city; (**g**) average criteria weights and (**h**) option preference of respondents from Kitakyushu city; (**i**) average criteria weights and (**j**) option preference of respondents from Moji; (**k**) average criteria weights and (**l**) option preference of respondents from Shimonoseki.

### 3.4. Fishers' Preference According to Fishing Methods and Scale

Fisheries preference was further analyzed due to their unique decision behavior in preferring the first option of sharing oceanographic information. Fisheries are the most prominent stakeholder group in local consensus building process as well as the most impacted local industry from the introduction of MRE projects. Interviews with fishery unions indicated that the impacts of the proposed options highly depend on their fishing methods, fishing grounds and scale. Figure 5 summarizes the fishery preferences based on fishing method and scale. In this analysis, grouping based on fishing method and fishing scale were highly inter-dependent. Most of the small- and medium-scale fishers were using the pole and line fishing method, whereas all respondent fishers who were grouped under the large-scale fishers category were using net fishing and longline fishing as the main fishing methods. Due to this equality of data sets, preference patterns of large-scale fishers and longline and net fishing fishers were exactly the same. Fishers usually use more than one fishing method. The most frequently used fishing method was considered for this grouping. Fishing method was significantly dependent on the fishing area. Most of the local fishers in Fukue and Naru Islands were small-scale fishers mainly using pole and line fishing. Fishers in Kitakyushu area mostly used large-scale fishing methods such as bottom draw nets and set nets.

From the interviews with fishers, we identified that the small-scale fishers who use pole and line method or nearshore fishing methods, such as diving, could benefit from the artificial reef effect and the fish gathering effect created by the subsea MRE structures. Hence, they preferred to have many small-scale MRE devices or structures in the area rather than a few large-scale MRE devices or structures. However, they acknowledged the technical factors that developers have to consider when designing the MRE device layout. In contrast to small-scale fishers, large-scale fishers who use fishing methods which need a large sea area to operate like longline, trawling and net fishing, prefer to have the least amount of MRE devices to minimize their fishing effort displacement. Since they use large sea area, real-time oceanographic information is vital to decide the travel plans and fishing grounds. Finally, large-scale fishers tend to be financially stronger than the small-scale fishers. Hence, large-scale fishers are more focused on the continuity of the industry and less willing to change the current practices, whereas small-scale fishers tended to prioritize different alternatives that provide more financial incentives. Interviews with fishers on Naru Island revealed that they prefer the benefit of having under water structures to support their fishing gears, specifically in the areas with strong tidal currents because, currently, they cannot use their fishing gear most of time due to the high tidal current velocity.

All these factors identified from key stakeholder interviews support the fisheries' preference shown in Figure 5. Large-scale fishers tended to prioritize the second criterion, environmental impacts, whereas small-scale fishers tended to prioritize economic impacts. This behavior can be explained by the current financial stability of the particular fishery groups. Generally, all the fishery groups tended to prefer the first option of sharing oceanographic information. However, in contrast with the large-scale fishers who use longline and net fishing methods, small- and medium-scale fishers who mostly use pole and line fishing indicated a significant preference for the second option of using MRE structures as artificial reefs and support structures for fishing gear.

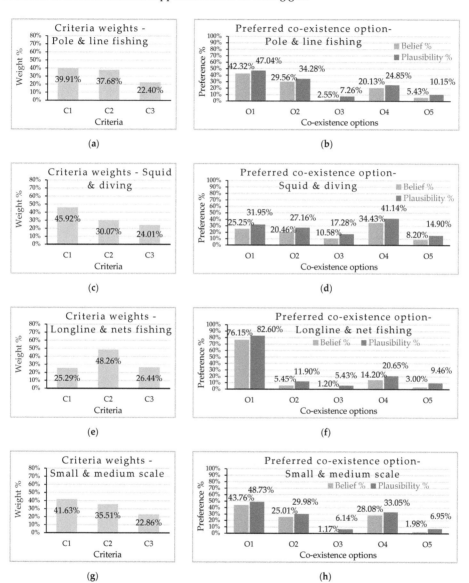

Figure 5. Cont.

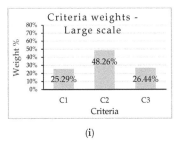 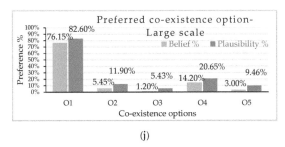

(i)                                                      (j)

**Figure 5.** Fishers' option preference according to fishing method and scale: (**a**) average criteria weights and (**b**) option preference of pole and line fishing fishers; (**c**) average criteria weights and (**d**) option preference of squid fishing and diving fishers; (**e**) average criteria weights and (**f**) option preference of longline and net fishing fishers; (**g**) average criteria weights and (**h**) option preference of small and medium scale fishers; (**i**) average criteria weights and (**j**) option preference of large scale fishers.

## 4. Discussion

The co-existence of MRE projects with existing marine activities is vital for building consensus among local stakeholders, which is essential for Japan's MRE developers. Understanding the compatibilities and co-location opportunities of different sea uses can lead to an optimum use of available sea area. Consensus building starts with identifying potential feasible negotiation options to create a win-win situation among the stakeholders and then identifying the different stakeholders' preferences for each option. In the case of MRE acceptance, key decision factors are as follows.

### 4.1. Nearshore vs. Offshore Projects

Nearshore MRE projects are more visible to the coastal communities. Hence, interactions between them are common. There is limited ocean space available in the nearshore area for specific marine activities such as shipping and transportation, docking fishery, and other commercial vessels. However, if the visual impacts are considered to be positive and used in a co-existence option, such as in the case of third option in this study, nearshore MRE projects are more preferred than the offshore MRE projects. Going further offshore can enhance the power takeout of the power plant, but increases the construction and operation costs. However, going offshore can create less congestion in the most competitive nearshore areas.

### 4.2. Different Technologies within MRE (Offshore Wind vs. Tidal Devices)

The co-existence strategy is dependent on the MRE technology. Certain types of co-existence options are compatible only with certain types of MRE projects. For example, providing oceanographic information is feasible with most of the MRE projects; however, the quality of data and the available parameters might be different from offshore wind project in comparison to tidal energy projects with fully submerged devices. The second option in this study, using MRE structures as artificial reefs and support structures for fishing gear, is more compatible with offshore wind energy projects than the tidal energy projects. However, tidal energy projects generally create less spatial conflict with fisheries because local fisheries generally do not use strong tidal current areas for fishing activities. Tidal energy projects require different skills and equipment, even for routine maintenance checks. However, some co-existence options can be used independent of the MRE technology used, such as the fourth option in this study—sharing generated electricity.

### 4.3. Knowledge, Perceptions and Values vs. Option Preference

Acceptance and preference depend on the decision makers' knowledge about the context and the perceived impacts of different decision alternatives. Some social science literature highlights the impact of personal values and beliefs on acceptance decision [60]. The best examples of this from the results of this study are the health and welfare group decision (Figure 3j) and the group decision of the Moji area residents. (Figure 4j). The health and welfare group perceived noise pollution and prioritized the environmental impacts over economic or social impacts and finally preferred the fourth option of having subsidized electricity. Most of those respondents have experienced nearby onshore wind turbines. The significant concerns about the low frequency noise pollution of offshore wind turbines indicate that they have a different belief about offshore wind turbines in comparison to onshore wind turbines. In contrast, Moji respondents prioritized economic impacts over environmental or social impacts, but still preferred the fourth option, which indicates that preference is highly correlated with personal economic gains.

*4.4. Cost of Co-Existence Options vs. Monetary Compensation Schemes*

The levelized cost of energy (LCOE), which represents the costs of electricity for an MRE installation over an assumed financial life and duty cycle [3], is significantly impacted by the installation costs and operation and maintenance (O&M) costs [61,62]. Limited information exists about the LCOE of Japan's MRE projects. However, from the literature on the European MRE industry [56,63,64], the initial commercial MRE projects in Japan are estimated to have a significant LCOE. Confidence in the ability of the MRE industry to deliver a competitive LCOE in comparison to other forms of power generation in an acceptable timeframe is essential for continued investment in the sector [65]. Hence, project developers were deeply concerned about managing the project costs to maintain a competitive LCOE with respect to other energy generation options. Even though the main focus of this study was not the analysis of the LCOE of MRE projects, it was important to analyze the potential economic impacts of the proposed co-existence strategies, since those options may significantly impact the LCOE and overall economic sustainability of the projects. The MCA results, in terms of the economic impacts (Table 4), show that stakeholders expect a positive overall impact from the proposed co-existence options except for the fourth option of sharing generated electricity at a subsidized rate. LCOE dynamics for various renewables depend on various factors that could be directly impacted by employing the identified co-existence options. For example, project developers may have to incur additional construction costs if options O1, O2, or O3 are employed. Conversely, using local resources according to O5 may reduce O&M costs. All these co-existence options may indirectly generate positive impacts on LCOE if they lead to a higher local acceptance level. However, there is not enough data to conduct an in-depth analysis of the effect of LCOE from individual co-existence strategies, which is a main limitation of this study.

An alternative to the identified non-monetary co-existence strategies is the monetary compensation or benefit creation scheme where the developers allocate funds to the local community with the mediation of local government, local authority, or some other responsible community body. Developers can benefit from the mediation of a local authority because they tend to have a higher trust among local communities and higher expertise in assessing the local priorities, which is vital when disbursing funds for the needs of the community. However, this kind of monetary benefit creation scheme worsens the developer costs and indirectly affects the LCOE. Developers can benefit if the LCOE impact of the proposed non-monetary co-existence strategies is worse than that of monetary benefit creation schemes. We suggest further studies to evaluate the economic impacts of each option.

## 5. Conclusions

A set of novel co-existence options that can be used to create a win-win situation among local coastal communities and the emerging MRE industry in Japan were evaluated based on DS-AHP in this study. Based on the key stakeholder interview results and a literature survey, the following co-existence options were considered: (1) providing real-time, in-situ ocean information from MRE farms; (2) using MRE structures as artificial reefs and support structures for fishing; (3) co-location with other industries such as leisure, tourism and aquaculture; (4) sharing generated electricity for local users at a subsidized rate; and (5) use of local resources to construct and operate the power plant, creating business involvement opportunities. By analyzing the preference results from the questionnaire survey, we found that stakeholders' decisions were mostly based on the perceived impacts on their daily lives. Local residents who interact less with the ocean generally preferred the stakeholder engagement aspect, with the intention of creating local benefits and thereby a win-win situation between the emerging MRE sector and traditional marine industries. Stakeholders who were interacting with the marine areas tended to know the real potential and limitations of the proposed co-existence options. Thus, there was a significant difference between the preference patterns of fishers, developers, and other stakeholders, where the final results indicate that fishers generally preferred the first option of sharing oceanographic information, whereas the general public and project developers preferred the fifth option of using local resources to construct and operate the power plant, thereby creating business involvement opportunities. Since these results were obtained by employing DS-AHP, the impact of uncertainty and data unavailability was minimal. We recommend using this approach not only when new information is available for the decision makers, for example after the initiation of real commercial MRE projects with more certain information and after commissioning the first commercial projects in the area, but also for other instances where stakeholder decisions have to be evaluated amidst a significant level of uncertainty and data unavailability.

**Author Contributions:** Conceptualization, A.H.T.S.K. and K.T.; methodology, A.H.T.S.K. and S.S.; software, A.H.T.S.K.; validation, A.H.T.S.K., and S.S.; formal analysis, A.H.T.S.K. and S.S.; investigation, K.T. and S.T.; resources, S.T. and K.T.; data curation, S.S.; writing—original draft preparation, A.H.T.S.K.; writing—review and editing, S.S., S.T. and K.T.; visualization, A.H.T.S.K.; supervision, K.T. and S.T.; project administration, A.H.T.S.K. and K.T.; funding acquisition, A.H.T.S.K. and K.T.

**Funding:** This research work was carried out with the financial support of Graduate Program in Sustainability Science-Global Leadership Initiative in the University of Tokyo. The APC was funded by Takagi Lab, Graduate school of frontier sciences, The University of Tokyo.

**Acknowledgments:** We are grateful to Hiroshi Matsuo of the Nagasaki Marine Industry Cluster Promotion Association (NaMICPA), Masanobu Shibuya of Shibuya diving industry Co., Ltd./Marine renewable energy-Fisheries co-existence center, and Hiroshi Ohza of the Goto city office for helping us to arrange the key stakeholder interviews. We are thankful to every key informant who participated our interviews. We are also thankful to Hirotaka Matsuda, Rukshani Liyanaarachchi, Norikazu Furukawa, Yuka Shimamura and Yang Jiaqi for the immense support given during the data collection and data analysis stages of this study. We are thankful to Onuki Motoharu, Atsuko Yamada, Yuko Opoku, Izumi Ikeda, and Naomi Sekine for helping with funding arrangements, travel arrangements, and questionnaire survey preparations. We also appreciate the editors and reviewers who gave valuable comments and suggestions to improve the manuscript.

**Conflicts of Interest:** The authors declare no conflict of interest. The funders had no role in the design of the study; in the collection, analyses, or interpretation of data; in the writing of the manuscript, or in the decision to publish the results.

## Abbreviations

| | |
|---|---|
| AHP | Analytic Hierarchy Process |
| BPA | Basic Probability Assignment |
| CMS | Condition Monitoring System (of the MRE power farm) |
| DST | Dempster Shafer Theory |
| DS-AHP | Dempster Shafer Analytic Hierarchy Process |
| EMEC | European Marine Energy Centre |
| GHG | Greenhouse Gas |
| LCOE | Levelized Cost Of Energy |
| MCA | Multi Criteria Analysis |
| MRE | Marine Renewable Energy |
| NaMICPA | Nagasaki Marine Industry Cluster Promotion Association |
| NPO | Non-Profit Organization |
| O&M | Operation and Maintenance |
| RIOE | Research Institute of Ocean Economics |

## References

1. Lewis, A.; Estefen, S.; Huckerby, J.; Musial, W.; Pontes, T.; Torres-Martinez, J. Ocean Energy. In *IPCC Special Report on Renewable Energy Sources and Climate Change Mitigation*; Edenhofer, O., Pichs-Madruga, R., Sokona, Y., Seyboth, K., Matschoss, P., Kadner, S., Zwickel, T., Eickemeier, P., Hansen, G., Schlömer, S., et al., Eds.; Cambridge University Press: Cambridge, UK; New York, NY, USA, 2011; pp. 497–534. Available online: https://www.ipcc.ch/site/assets/uploads/2018/03/Chapter-6-Ocean-Energy-1.pdf (accessed on 21 April 2019).
2. Alistair, G.L. Borthwick, Marine Renewable Energy Seascape. *Engineering* **2016**, *2*, 69–78. [CrossRef]
3. Soukissian, T.; Denaxa, D.; Karathanasi, F.; Prospathopoulos, A.; Sarantakos, K.; Iona, A.; Georgantas, K.; Mavrakos, S. Marine Renewable Energy in the Mediterranean Sea: Status and Perspectives. *Energies* **2017**, *10*, 1512. [CrossRef]
4. Huckerby, J.; Jeffrey, H.; Sedgwick, J.; Jay, B.; Finlay, L. An International Vision for Ocean Energy—Version II. 2012. Available online: http://www.policyandinnovationedinburgh.org/uploads/3/1/4/1/31417803/oes_booklet_fa_print_08_10_2012.pdf (accessed on 21 April 2019).
5. Appiott, J.; Dhanju, A.; Cicin-Sain, B. Encouraging renewable energy in the offshore environment. *Ocean Coast. Manag.* **2014**, *90*, 58–64. [CrossRef]
6. IEA Wind, Expert Group Summary on Recommended Practices—Social Acceptance of Wind Energy Projects. 2013. Available online: http://www.socialacceptance.ch/images/RP_14_Social_Acceptance_FINAL.pdf (accessed on 21 April 2019).
7. Wiersma, B.; Devine-Wright, P. Public engagement with offshore renewable energy: A critical review, Wiley Interdiscip. *Rev. Clim. Chang.* **2014**, *5*, 493–507. [CrossRef]
8. Japan Wind Power Association. *JWPA Report on Act on using marine areas for Marine Renewable Energy Projects*; Japan Wind Power Association: Tokyo, Japan, 2019; Available online: http://jwpa.jp/page_276_englishsite/jwpa/detail_e.html (accessed on 21 April 2019).
9. Prime Ministers Headquater for Ocean Policy in Japan, Outline of the Third Basic Plan on Ocean Policy, 2018. Available online: http://www8.cao.go.jp/ocean/english/plan/pdf/plan03_gaiyou_e.pdf (accessed on 21 April 2019).
10. Firestone, J.; Kempton, W. Public opinion about large offshore wind power: Underlying factors. *Energy Policy* **2007**, *35*, 1584–1598. [CrossRef]
11. Walker, B.J.A.; Wiersma, B.; Bailey, E. Community benefits, framing and the social acceptance of offshore wind farms: An experimental study in England. *Energy Res. Soc. Sci.* **2014**, *3*, 46–54. [CrossRef]
12. Cass, N.; Walker, G.; Devine-Wright, P. Good neighbours, public relations and bribes: The politics and perceptions of community benefit provision in renewable energy development in the UK. *J. Environ. Policy Plan.* **2010**, *12*, 255–275. [CrossRef]
13. Aitken, M. Wind power and community benefits: Challenges and opportunities. *Energy Policy* **2010**, *38*, 6066–6075. [CrossRef]
14. Walker, B.J.A.; Russel, D.; Kurz, T. Community Benefits or Community Bribes? An Experimental Analysis of Strategies for Managing Community Perceptions of Bribery Surrounding the Siting of Renewable Energy Projects. *Environ. Behav.* **2017**, *49*, 59–83. [CrossRef]
15. Reilly, K.; O'Hagan, A.M.; Dalton, G. Developing benefit schemes and financial compensation measures for fishermen impacted by marine renewable energy projects. *Energy Policy* **2016**, *97*, 161–170. [CrossRef]
16. Yates, K.L.; Schoeman, D.S. Spatial Access Priority Mapping (SAPM) with Fishers: A Quantitative GIS Method for Participatory Planning. *PLoS ONE* **2013**, *8*. [CrossRef]
17. RIOE (Research Institute for Ocean Economics—Japan). *Recommendations on Consensus Building with Fishery Cooperatives for Offshore Wind Power Projects*; RIOE: Chiba, Japan, 2013; Available online: http://www.rioe.or.jp/0510teigen.pdf (accessed on 21 April 2019).
18. Goto City Office, Goto Ocean Energy, (n.d.) 1–15. Available online: https://www.env.go.jp/nature/biodic/coralreefs/iccccrc2013/pdf/year2013630/section2/noguchi.pdf (accessed on 21 April 2019).
19. Nagasaki Marine Industry Cluster Promotion Association (NaMICPA), (n.d.). Available online: http://namicpa.jp/ (accessed on 7 August 2018).
20. Japan Wind Power Association. *Offshore Wind Power Development in Japan*; Japan Wind Power Association: Tokyo, Japan, 2017; 28p, Available online: http://jwpa.jp/pdf/20170228_OffshoreWindPower_inJapan_r1.pdf (accessed on 21 April 2019).

21. Waldman, S.; Yamaguchi, S.; Murray, R.O.; Woolf, D. Tidal resource and interactions between multiple channels in the Goto Islands, Japan. *Int. J. Mar. Energy* **2017**, *19*, 332–334. [CrossRef]
22. New Energy and Industrial Technology Development Organisation (NEDO). *NEDO Offshore Wind Energy Progress*, 2nd ed.; NEDO: Kawasaki City, Japan, 2013; 24p, Available online: http://www.nedo.go.jp/content/100534312.pdf?from=b (accessed on 21 April 2019).
23. Hibiki Wind Energy Co. Ltd. *Hibiki Wind Energy*; Hibiki Wind Energy Co. Ltd.: Kyushu, Japan, 2018; 4p, Available online: http://hibikiwindenergy.co.jp/pdf/hwe_english.pdf (accessed on 21 April 2019).
24. Saaty, R.W. The analytic hierarchy process—What it is and how it is used. *Math. Model.* **1987**, *9*, 161–176. [CrossRef]
25. Saaty, T.L. How to make a decision: The analytic hierarchy process. *Eur. J. Oper. Res.* **1990**, *48*, 9–26. [CrossRef]
26. Dempster, A. Upper and lower probabilities induced by a multi-valued mapping. *Ann. Math. Stat.* **1967**, *38*, 325–339. [CrossRef]
27. Dempster, A. A Generalization of Bayesian Inference. *J. R. Stat. Soc. Ser. B* **1968**, *30*, 205–232. [CrossRef]
28. Shafer, G. *A Mathematical Theory of Evidence*; Princeton University Press: Princeton, NJ, USA, 1976.
29. Beynon, M.J.; Cosker, D.; Marshall, D. An expert system for multi-criteria decision making using Dempster Shafer theory. *Expert Syst. Appl.* **2001**, *20*, 357–367. [CrossRef]
30. Beynon, M.J. DS/AHP method: A mathematical analysis, including an understanding of uncertainty. *Eur. J. Oper. Res.* **2002**, *140*, 148–164. [CrossRef]
31. Beynon, M.J. A method of aggregation in DS/AHP for group decision-making with the non-equivalent importance of individuals in the group. *Comput. Oper. Res.* **2005**, *32*, 1881–1896. [CrossRef]
32. Awasthi, A.; Chauhan, S.S. Using AHP and Dempster–Shafer theory for evaluating sustainable transport solutions. *Environ. Model. Softw.* **2011**, *26*, 787–796. [CrossRef]
33. Kinoshita, K. Evaluation of Regional Characteristics of Wave Energy and Research of Mooring System. 2010. Available online: https://www.env.go.jp/earth/ondanka/cpttv_funds/pdf/prod20100301.pdf (accessed on 21 April 2019).
34. Fischer, J.; Flemming, N. *Operational Oceanography: Data Requirements Survey*; EuroGOOS Publication No. 12; Southampton Oceanography Centre: Southampton, UK, 1999; ISBN 0-904175-36-7. Available online: https://www.academia.edu/21168082/The_EuroGOOS_data_requirements_survey (accessed on 11 May 2019).
35. Chiabai, A.; Nunes, P.A.L.D. Economic Valuation of Oceanographic Forecasting Services: A Cost-Benefit Exercise, 2006. Available online: http://www.feem.it/userfiles/attach/Publication/NDL2006/NDL2006-104.pdf (accessed on 21 April 2019).
36. Francisco, F.; Sundberg, J. Detection of Visual Signatures of Marine Mammals and Fish within Marine Renewable Energy Farms using Multibeam Imaging Sonar. *J. Mar. Sci. Eng.* **2019**, *7*, 22. [CrossRef]
37. Hooper, T.; Austen, M. The co-location of offshore windfarms and decapod fisheries in the UK: Constraints and opportunities. *Mar. Policy* **2014**, *43*, 295–300. [CrossRef]
38. Hooper, T.; Ashley, M.; Austen, M. Perceptions of fishers and developers on the co-location of offshore wind farms and decapod fisheries in the UK. *Mar. Policy* **2015**, *61*, 16–22. [CrossRef]
39. de Groot, J.; Campbell, M.; Ashley, M.; Rodwell, L. Investigating the co-existence of fisheries and offshore renewable energy in the UK: Identification of a mitigation agenda for fishing effort displacement. *Ocean Coast. Manag.* **2014**, *102*, 7–18. [CrossRef]
40. Stelzenmüller, V.; Diekmann, R.; Bastardie, F.; Schulze, T.; Berkenhagen, J. Co-location of passive gear fisheries in offshore wind farms in the German EEZ of the North Sea: A first socio-economic scoping. *J. Environ. Manag.* **2016**, *183*, 794–805. [CrossRef]
41. Blyth-Skyrme, R. *Benefits and Disadvantages of Co-Locating Windfarms and Marine Conservation Zones*; Report to Collaborative Offshore Wind Research into the Environment Ltd.: London, UK, 2010; p. 37. Available online: https://tethys.pnnl.gov/sites/default/files/publications/Blyth-Skyrme-2011.pdf (accessed on 21 April 2019).
42. Gimpel, A.; Stelzenmuller, V.; Grote, B.; Buck, B.H.; Floeter, J.; Nunez-Riboni, I.; Pogoda, B.; Temming, A. A GIS modelling framework to evaluate marine spatial planning scenarios: Co-location of offshore wind farms and aquaculture in the German EEZ. *Mar. Policy* **2015**, *55*, 102–115. [CrossRef]

43. Mackinson, S.; Curtis, H.; Brown, R.; Mctaggart, K.; Taylor, N.; Neville, S.; Rogers, S. A report on the perceptions of the fishing industry into the potential socio-economic impacts of offshore wind energy developments on their work patterns and income. *Sci. Ser. Tech. Rep. Cefas Lowestoft* **2006**, *133*, 99. Available online: https://tethys.pnnl.gov/sites/default/files/publications/Perceptions_of_the_Fishing_Industry_on_Offshore_Wind.pdf (accessed on 21 April 2019).
44. Buck, B.H.; Krause, G.; Rosenthal, H. Extensive open ocean aquaculture development within wind farms in Germany: The prospect of offshore co-management and legal constraints. *Ocean Coast. Manag.* **2004**, *47*, 95–122. [CrossRef]
45. Buck, B.H.; Langan, R. *Aquaculture Perspective of Multi-Use Sites in the Open Ocean*; Springer: Berlin, Germany, 2017. [CrossRef]
46. Wang, L. Comparative Study of Wind Turbine Placement Methods for Flat Wind Farm Layout Optimization with Irregular Boundary. *Appl. Sci.* **2019**, *9*, 639. [CrossRef]
47. Fayram, A.H.; de Risi, A. The potential compatibility of offshore wind power and fisheries: An example using bluefin tuna in the Adriatic Sea. *Ocean Coast. Manag.* **2007**, *50*, 597–605. [CrossRef]
48. Westerberg, V.; Jacobsen, J.B.; Lifran, R. The case for offshore wind farms, artificial reefs and sustainable tourism in the French mediterranean. *Tour. Manag.* **2013**, *34*, 172–183. [CrossRef]
49. RYA & CA. 'Sharing the Wind' Recreational Boating in the Offshore Wind Farm Strategic Areas, 2004. Available online: https://www.rya.org.uk/sitecollectiondocuments/legal/WebDocuments/Environment/SharingtheWindcompressed.pdf (accessed on 21 April 2019).
50. MAFF (Ministry of Agriculture Forestry and Fisheries). *Japan, Promotion Project of Eco-Friendly Fishing Ports*; MAFF: Tokyo, Japan, 2012. Available online: http://www.maff.go.jp/j/aid/hozyo/2012/suisan/pdf/80.pdf (accessed on 21 April 2019).
51. Cascajo, R.; García, E.; Quiles, E.; Correcher, A.; Morant, F. Integration of Marine Wave Energy Converters into Seaports: A Case Study in the Port of Valencia. *Energies* **2019**, *12*, 787. [CrossRef]
52. Clayton, M.; Stillwell, A.; Webber, M. Implementation of Brackish Groundwater Desalination Using Wind-Generated Electricity: A Case Study of the Energy-Water Nexus in Texas. *Sustainability* **2014**, *6*, 758–778. [CrossRef]
53. FLOWW Fishing Liaison with Offshore Wind and Wet Renewables Group. Recommendations for Fisheries Liaison, 2008. Available online: https://webarchive.nationalarchives.gov.uk/+/http:/www.berr.gov.uk/files/file46366.pdf (accessed on 21 April 2019).
54. Seyr, H.; Muskulus, M. Decision Support Models for Operations and Maintenance for Offshore Wind Farms: A Review. *Appl. Sci.* **2019**, *9*, 278. [CrossRef]
55. Nachimuthu, S.; Zuo, M.J.; Ding, Y. A Decision-making Model for Corrective Maintenance of Offshore Wind Turbines Considering Uncertainties. *Energies* **2019**, *12*, 1408. [CrossRef]
56. OES—IEA. International Levelised Cost of Energy for Ocean Energy Technologies, 2015. p. 35. Available online: http://www.ocean-energy-systems.org/news/international-lcoe-for-ocean-energy-technology/?source=newsletter (accessed on 21 April 2019).
57. Gao, X.; Xia, L.; Lu, L.; Li, Y. Analysis of Hong Kong's Wind Energy: Power Potential, Development Constraints, and Experiences from Other Countries for Local Wind Energy Promotion Strategies. *Sustainability* **2019**, *11*, 924. [CrossRef]
58. Ishidoya, H. Studies on Kyucho Events and Disaster Prevention of Set Nets in Sagami Bay. *Fish. Sci.* **2002**, *68*, 1841–1844. [CrossRef]
59. Matsuyama, M.; Ishidoya, H.; Iwata, S.; Kitade, Y.; Nagamatsu, H. Kyucho induced by intrusion of Kuroshio water in Sagami Bay, Japan. *Cont. Shelf Res.* **1999**, *19*, 1561–1575. [CrossRef]
60. Bidwell, D. Ocean beliefs and support for an offshore wind energy project. *Ocean Coast. Manag.* **2017**, *146*, 99–108. [CrossRef]
61. Myhr, A.; Bjerkseter, C.; Ågotnes, A.; Nygaard, T.A. Levelised cost of energy for offshore floating wind turbines in a lifecycle perspective. *Renew. Energy* **2014**, *66*, 714–728. [CrossRef]
62. Allan, G.; Gilmartin, M.; McGregor, P.; Swales, K. Levelised Costs of Wave and Tidal Energy in the UK: Cost Competitiveness and the Importance of "Banded" Renewables Obligation Certificates. *Energy Policy* **2011**, *39*, 23–39. [CrossRef]

63. Astariz, S.; Vazquez, A.; Iglesias, G. Evaluation and comparison of the levelized cost of tidal, wave, and offshore wind energy. *J. Renew. Sustain. Energy* **2015**, *7*, 053112. [CrossRef]
64. Lerch, M.; De-Prada-Gil, M.; Molins, C.; Benveniste, G. Sensitivity analysis on the levelized cost of energy for floating offshore wind farms. *Sustain. Energy Technol. Assess.* **2018**, *30*, 77–90. [CrossRef]
65. Weller, S.; Thies, P.; Gordelier, T.; Johanning, L. Reducing Reliability Uncertainties for Marine Renewable Energy. *J. Mar. Sci. Eng.* **2015**, *3*, 1349–1361. [CrossRef]

© 2019 by the authors. Licensee MDPI, Basel, Switzerland. This article is an open access article distributed under the terms and conditions of the Creative Commons Attribution (CC BY) license (http://creativecommons.org/licenses/by/4.0/).

*Article*

# Revising the Environmental Kuznets Curve for Deforestation: An Empirical Study for Bulgaria

Stavros Tsiantikoudis [1,*], Eleni Zafeiriou [2], Grigorios Kyriakopoulos [3] and Garyfallos Arabatzis [1]

[1] Department of Forestry and Management of the Environment and Natural Resources, Democritus University of Thrace, GR68200 Orestiada, Greece
[2] Department of Agricultural Development, Democritus University of Thrace, GR68200 Orestiada, Greece
[3] School of Electrical and Computer Engineering, National Technical University of Athens (NTUA), Division of Electric Power, GR15780 Athens, Greece
\* Correspondence: stsianti@fmenr.duth.gr

Received: 24 June 2019; Accepted: 8 August 2019; Published: 12 August 2019

**Abstract:** The evolution of human societies along with efforts to enhance economic welfare may well lead to the deterioration of the environment. Deforestation is a usual process throughout evolution that poses pressing and potentially irreversible environmental risks, despite the ecological and modernization processes that aim to limit those risks. The economic growth–environmental degradation relationship—namely, the environmental Kuznets curve (EKC) hypothesis—is studied in alignment with the autoregressive distributed lag (ARDL) approach. The novelty of the study is attributed to the use of the carbon emissions equivalent derived by deforestation as an index for environmental degradation in Bulgaria as a new entrant into the European Union (EU). In addition, we use the gross domestic product (GDP) per capita as a proxy for income, being determined as an independent variable. Research findings cannot validate the inverted U-shape of the EKC hypothesis; instead, an inverted N pattern is confirmed. The implementation of appropriate policies aiming at the protection of the environment through the diversification of economic activities is related to the use of forest land and other resources, or related sectors (agroforestry, ecotourism activities, and scientific research), rather than only the direct utilization of forested areas; the limitation of afforestation processes and their negative impacts on citizens' welfare are also addressed.

**Keywords:** environmental Kuznets curve; deforestation; ARDL with bounds test

## 1. Introduction

The multifunctionality of forests in Europe stems from the diversity of tree species and ecosystems. Indicatively, forests serve as a regulatory tool for hydrologic cycles, provide refuges for biodiversity, contribute raw material for medicinal and forest products, prevent soil destruction, and satisfy recreational, spiritual, and aesthetic value needs [1]. However, the most significant impact is related to climate change mitigation, since it contributes greatly to exchanges in energy, water, carbon dioxide, and other chemical substances [1,2]. The significant contribution of forests to climate change mitigation has become recognized in the existing literature recently [3]. Given that the destruction of forests reduces the ability of the Earth to absorb $CO_2$ from the atmosphere [4], an effort was initiated for its limitation. This effort was established with the negotiation of the Montreal UNFCCC CoP held in 2005, in which the dominant motto for forest management is "Reducing emissions from deforestation and forest degradation" (REDD).

In Europe, forests correspond to 37% of the terrestrial surface with ecological, economical, and soci-ocultural impacts, as mentioned above [5,6], while eastern and central European countries do still have large and relatively undisturbed forests compared to western European countries [7,8].

Within the last couple of centuries in the name of societal modernization and urbanization, deforestation has become a necessary tool, which permanently converts forest land into other land uses [9].

A high level of deforestation is globally considered an anthropogenic environmental problem [9,10]. The impacts of this problem are more severe for less developed countries, and less extended for developed countries. Indeed, developed countries are exerting intense efforts for afforestation, which led to an increase in the forest area, estimated at 1% from 1990 to 2005 [11]. The linkage among deforestation, forest degradation, and climate change is attributed to fewer trees absorbing less greenhouse gases (GHGs); therefore, deforestation is indirectly causing increased quantities of carbon dioxide emissions [12,13]. In addition, according to macro-sociologists, carbon dioxide emissions may be indicative of marketable outsourcing production for developing countries [13]. Deforestation in global terms is related to the cooling effect and the warming carbon cycle, due to changes in albedo and evapotranspiration [14].

Deforestation in the European Union (EU) constitutes an environmental issue of adverse negative impacts within the last few decades. Motivation policies for deforestation are characterized by complexity and are differentiated from region to region and from country to country. One of the major reasons for deforestation in the EU, according to the FAO (2016) [15], is related to the agricultural expansion for the production of specific food commodities (accounting for 80%), while the issues of urbanization/infrastructure interpret less than 10% of deforestation each.

Bulgaria is selected as the country to be studied, since it has become a new member state in the EU in 2007 and it is ranked fourth among EU members in terms of gross domestic product (GDP) growth for the last decade. The change from a highly centralized, planned economy to an open, market-based, upper-middle income economy has become achievable after a decade of slow economic restructuring and growth, high indebtedness, and a loss of savings. The time period until its entrance into the EU was characterized by an exceptionally high economic growth and improved living standards. However, in the last decade, few positive impacts were implemented. Particularly, the global economic crisis of 2008 and a period of political instability in the period 2013–2014 urged the Bulgarian government to achieve the objectives of growth and shared prosperity.

Today, for Bulgaria, the most importance issues to be addressed are those of raising productivity and handling the rapid demographic change. Particularly, higher productivity growth is critical to accelerate convergence, as Bulgaria's income per capita is only 47% of the EU average, which is the lowest in the EU.

Furthermore, in Bulgaria, according to statistics provided by the FAO [15], 36.1% or about 3,927,000 ha of the total area is forested, 8.6% (338,000.00) of which is classified as primary forest, which is the mostly bioversatile and carbon-dense forest area. In addition, 815,000 ha are forest plantations. Besides, deforestation has currently become a significant issue, since Bulgaria's forests contain 202 million metric tons of carbon in living forest biomass.

The overexploitation of forests—being not accompanied by a project for sustainable management—as well as a lack of effective surveillance of forests and forest areas in Bulgaria resulted in losing an average of 30,000 ha or 0.90% per year [16].

Common methods of deforestation for Bulgaria are burning trees and clear cutting. These practices are considered controversial, since they leave the land completely barren. Degradation and deforestation may well lead to a devastating chain of events both locally and globally, including the loss of species, the water cycle (trees are important to the water cycle, since they absorb rain fall and produce water vapor that is released into the atmosphere), soil erosion, life quality, and floods during winter.

The low income is encouraging the government to implement institutional and legislative changes, such as low tax rates and promotional investments. On the other hand, the country's forest cover has been expanding gradually, at an annual rate of 0.6% over the period 1990–2015 [17], while as an EU member, Bulgaria is required to implement the EU Timber Regulation, which came into force in March 2013.

Another major alarming problem with the forests in Bulgaria is illegal logging. Among the types of illegal logging are the illegal lending of forest area and damage of forest stands to obtain extended harvesting volume at a lower price.

Nowadays, the legislation has become strict regarding the activities and rights related to forest management and timber harvesting. The state forestry staff should control logging activities, but all the violations in forests are caused by a lack of governance and enforcement of existing regulations and laws, thus necessitating cooperation among forest employees. Moreover, low salaries have led to high rates of corruption in terms of state forest services as well as to limited and ineffective control mechanisms [16]. It is noteworthy that deforestation and forest degradation account for approximately 20–25% of global anthropogenic greenhouse gas (GHG) emissions, which are the major source of emissions from developing countries [18]. Deforestation in Bulgaria, as among all ex-socialist countries, and especially in private forests, is becoming limited according to statistical data, which is attributed to the modernization, urbanization, and immigration phenomena that were initiated in the period of socialism. In Bulgaria, data are available only for the case of afforestation, but not for deforestation. An indirect conclusion can be reached through the change in the land area covered by forests as well as the afforestation rate. Particularly, a decrease in the afforestation rate was observed during the 1990s, when the afforested areas per year were decreased under 10,000 ha year$^{-1}$, and after 2009, under 5000 ha year$^{-1}$. The major reasons for this decrease are the following: the large mastered areas for afforestation, the development of silvicultural systems with natural regeneration priority, chronic economic crises, as well as a lack of resources.

There are plentiful studies addressing the environmental Kuznets curve (EKC) hypothesis, where deforestation has been used as a proxy for environmental degradation, given the direct effects of growth on natural capital, in order productive agrarian efforts and policies to be promoted [19]. A lack of data for the deforestation area was the reason that the authors in this study used the GHG emissions generated by deforestation as an index for environmental degradation.

In this research context, the extended deforestation and the carbon emissions generated by deforestation are related to the GDP per capita (as motivation for income improvement). This scientific background can be a subject of econometric study under the framework of the EKC on providing policy tools to enable a strategy design that can provide alternative and more profitable sources of income. In this study, the ARDL bounds cointegration technique was deployed, and it validated the reversed N Kuznets pattern for the data applied. The novelty of the study stands on the use of the GHG emissions generated by deforestation for a country with many particularities regarding the issue of deforestation, including a lack of data, suffering from the problem of illegal logging, and strong motivation for afforestation. The study is organized as follows: Section 2 describes the existing literature, Section 3 outlines the methodology, Section 4 provides and discusses the results, and the concluding remarks are succinctly presented in Section 5.

## 2. Literature Overview

### 2.1. Environmental Kuznets Curve Hypothesis (EKC)

The EKC hypothesis is an empirical relationship that assumes the existence of a relationship between environmental quality or pollutant emissions and economic growth. Two plausible explanations have been suggested for the interpretation of the environmental–economic performance relationship [20,21]. The first interpretation is related to an income effect, because the environment is valued as a luxurious asset. Therefore, in the initial stages of the economic development process, the individuals are not willing to trade consumption for investment in environmental protection, resulting in a decline in environmental quality. Once individuals reach a specific level of consumption, which is known in the EKC literature as the "income turning point", they ask for increasing investments for improving the environment. In addition, after the turning point, environmental quality indicators (showing pollution and environmental degradation) begin to improve.

Potential factors for the interpretation of the EKC are the following: first, the scale of production related to production expansion with the mix of products produced, and the mix of production inputs used, is considered constant under the condition of technological status. Second, different industries are characterized by different pollution intensities, while the output mix typically changes along with economic development. Third, changes in the input mix may well be related to the substitution of less environmentally damaging inputs to production for more damaging inputs, and vice versa. Another significant factor is the trade openness and specifically the increasing trade openness in the case of Tunisia, which is validated as statistically significant with both linear and nonlinear ARDL [22]. Four, it is noteworthy that the improvements in the state of technology, especially those changes in production efficiency and emissions-specific changes in process that result in less pollutant emitted per unit of input.

Nevertheless, there has been not a consensus on the validation or rejection of the EKC hypothesis. Specifically, several studies are devoted to environmental degradation as an endogenous variable and income per capita as an exogenous variable with mixed results [23].

## 2.2. Previous Studies

### 2.2.1. EKC General

Starting from Grossman and Krueger [24], plentiful studies can be mentioned with differences in terms of study period, methodological adaptation, power of income, and choice of control variables [25]. The data employed can be time series cross-section, or even panel data, also with different results [26,27]. For the time period 1991–2009, the time-series analysis involved nearly eight broad categories of methods, while the results obtained from these studies are inconclusive. An inverted U-shape was validated for France by Ang [28] for 1960–2000 with the assistance of ARDL methodology; Ozatac et al. also validated an EKC pattern for Turkey [29] with the same methodology for the period from 1960 to 2013. On the other hand, the N-shape EKC pattern was a finding by Akbostancı et al. [30] for Turkey for the period from 1968 to 2003, with the assistance of the cointegration technique. Regarding the panel data, the most common methodology is FMOLS, which Apergis and Payne introduced [26] for the study of six Central American countries for the period of 1971 to 2004, providing evidence of an inverted U-shaped EKC. The same conclusion was obtained by Liu et al. [31] referring to 10 newly industrialized countries for the period from 1971 to 2013, and Aruga (2019) [27] for a number of Asian Pacific countries.

### 2.2.2. EKC and Deforestation

Deforestation is a process that is increasing as a result of economic expansion. Therefore, interaction among deforestation and prosperity has become a subject of extended studies with significant efforts to be succinctly described below.

Cropper and Griffiths [32] valued deforestation not as an environmental degradation index, but as an environmental management measure. Specifically, this study involved the impact of population and income increase on the reduction of forest areas for 64 developing countries including those in Africa, Asia, and Latin America, which all feature huge forests and forest areas. According to these findings, the EKC was validated for countries in Latin America and Africa, where the per-capita GDP is lower than the estimated highest point of the curve ($5.420 and $4760 respectively in 1985 prices or at about $9100 and $7900 in 2001 prices. Therefore, the ECK did not reach the highest point, implying limitations in the validation of the EKC.

The constant growth of the global population and the adverse exploitation of natural sources to satisfy human needs, have led to deforestation along with conversion to agricultural lands. The depletion of the world's forests in both tropical and temperate regions is causing considerable environmental problems that hamper sustainable economic development. In approaching the deforestation trend, some researchers argued that this activity might be slow or reverse, unveiling the validity of the EKC hypothesis. Nevertheless, the results and conclusions of studies investigating EKC contradict each other. Moreover, the relevant literature compared OECD countries with the non-OECD countries of Latin America, Asia, and Africa in determining the ways under which the various factors of economic growth, population, trade, urbanization, agricultural land conversion, and cereal yield impact deforestation rates [33]. These authors concluded that the OECD countries present an N-shaped curve, while for the African region, an income-based EKC pattern is validated. It is also noteworthy that the trade openness and the internationalization of the marketplace, along with widespread urbanization, are all impacting the regions in a different way, but only these countries have shown less deforestation attributed to higher cereal yields [33].

The validity and credibility of the EKC method to examine the deforestation process has attracted plentiful studies. Even though deforestation is widely studied, the relationship between economic development and deforestation remains questionable. In this context, a meta-analysis of selected EKC studies for deforestation included 69 studies, offering 547 estimations regarding the differentiation and vulnerability of EKC-reported results [34]. These authors investigated a range of choices—such as econometric strategy, measures of deforestation, geographical area, and the presence of control variables—on the probability of finding an EKC. It was argued that the validity and credibility of the EKC method is solidified, and the theoretical alternatives to the predominance of EKC as a method to examine the deforestation process could fade [34]. In deepening scientific understanding upon the functionality of EKC and in perceiving alternative means of explaining how capitalism and ecological disorganization (pollution) are interconnected under Marxist theory, Lynch [35] illustrated the limited ability of common EKC interpretation and attempted to implement a Marxian interpretation of the EKC. Under this context, Marxian analysis supports a reasonable interpretation for the inverted "U" shape of pollution where increased pollution is expected due to ongoing economic development. Furthermore, occasionally, the orthodox EKC may fail to fulfill EKC predictions in developing countries, thus implying that pollution rises in these areas. In the long term, pollution is reduced in poor nations, but it leads to a rise in pollution in the global context. However, as time passes, the production in developing economies is increasing, which leads in turn to an upward trend for pollution [35].

The main socio-environmental parameters that determine deforestation are the: property rights, the agricultural price index, the forest area, population, income, and timber prices [36]. Therefore, a plausible methodological tool to investigate the impact of deforestation-induced factors is the EKC. Under this research context, Esmaeili and Nasrnia [36] deployed the autoregressive distributed lag approach to yield the deforestation function, confirming the existence of an inverted U-shaped EKC for deforestation in Iran. Moreover, their research proved a linkage among deforestation and property rights, forest area, agricultural price index, and terms of trade. Additionally, the key aspects that support sustainable forest land uses and a reduction in the deforestation rate are the improvement of secure property rights and the environmental policies planning in Iran [36].

Regarding the roles of trade openness and agricultural productivity under the EKC, researchers have investigated these factors under the applicability of the EKC hypothesis in relation to emissions from the non-oil sector, such as agriculture (i.e., all GHG emissions from agriculture) among the sub-Saharan African countries [37]. Particularly, authors were focused on two indicators of environmental change, namely that of "rate of deforestation" and "all greenhouse gas emissions from agriculture" (Agri-GHG) in addressing whether the EKC hypothesis exists for both indicators applied and, secondly, to investigate the effects of macroeconomic and institutional variables on both of these previously considered indicators. Specifically, it is noteworthy that: (a) the EKC exists (such as an inverted "U"-shaped) only for all GHG emissions from agriculture; (b) agricultural production and trade openness significantly increase both of the applied environmental change indicators, and (c) population growth significantly reduces Agri-GHG, while economic growth significantly increases the "rate of deforestation" in the region examined [37].

Zafeiriou et al. [38] studied the temporal environmental degradation agricultural income relationship for three ex-socialist newly entrants in the EU (Bulgaria, the Czech Republic, and

Hungary), for which the EKC hypothesis was confirmed for the first two countries, while for Hungary, it was not confirmed upon the specific data employed.

Barbier and Burgess [39] focused on a study of deforestation estimation in tropics through the extension of cultivated areas. The sample involves countries of the tropics in Africa, Latin America, and Asia for the period from 1961 to 1994. According to their findings, the EKC was validated as a turning point, generating $8700.00 GDP per capita in 2003 prices. The model included a number of economic and social variables, such as GDP growth, the population increase rate, cereal production, land-use distribution, the exports of agricultural products, political corruption, and political stability, as well as property rights implementation. According to these findings, the expansion of agricultural area was positively affected by the population increase rate, while the agricultural land distribution, exports, and political stability positively impacted the dependent variable [39].

Similarly, Ehrhardt-Martinez et al. [40] utilized data for the time period from 1980 to 1995 and for the 74 least developed countries in Africa, Asia, and Latin America. These authors studied the validity of the ECK hypothesis and signified that the threshold value is estimated at $1150 in 1980 prices (or $2354.00 in 2003 prices).

In another study regarding deforestation, Lantz [41] used the annually deforestation rates in five Canadian regions for the time period of 1975 to 1999 as proxy for forest area deforestation in order to unveil different relationship patterns, including those between income–deforestation, population–deforestation, and technological improvements–deforestation, across time. According to these findings, the deforestation area was negatively related to income evolution, which was constantly decreasing. Therefore, the impacts on the forest destruction lead to higher incomes, and they are more efficiently reflected to the population [41].

Chiu [42] validated that deforestation was decreased in alignment with the income increase in more than 52 developing countries. Similarly, Farhani et al. [43] confirmed the validity of EKC for the relationship between income–environmental sustainability in South African and Middle Eastern countries. The study of Parajuli et al. (2019) [44], who studied the impact of forests, agricultural area, and energy consumption on the carbon emissions generated by 86 countries by developing the dynamic panel data approach, is also noteworthy.

2.2.3. EKC Methodological Issues

The validity of the environmental degradation–income per capita relationship has been subjected to plentiful studies within the last two decades, which is in alignment with the different indexes considered, and also with different linear and nonlinear methodologies applied [38,45–52]. The deterioration of the environmental problems caused by carbon emissions and especially GHG emissions, the identification of the impacts of climate change on global economy [53], and the concepts of eco-efficiency sustainability under the context of the global economy supported motives for these theoretically approached scientific fields. The prevailing econometric methodology upon the environmental degradation–economic growth relationship is either linear or nonlinear cointegration. The inverted U pattern of this relationship was initially introduced by Grossman and Krueger [47]. Esteve and Tamarit [54] outlined the most significant empirical studies, including those performed by Ozturk and Acaravci [55], Halicioglou [56], Soytas and Sari [57], and Soytas et al. [23], showing contradicting results with linearity to be the common feature among all of them. The existence of nonlinearities was initially studied by Esteve and Tamarit [54], noting that the implementation of the threshold cointegration validated the EKC hypothesis for Spain over a long-run reference period.

The present study, with the aid of the ARDL cointegration technique, investigated the relationship of environmental degradation–economic growth by using the: (a) carbon emissions generated due to deforestation as an index for environmental degradation, and (b) GDP per capita as an index for economic growth for the new EU entrant country of Bulgaria. The use of this index for environmental degradation is the novel contribution of this study, comparing to existing literature.

## 3. Data Methodology

### 3.1. Data

In the study, authors employed carbon emissions generated by deforestation as an index for environmental degradation for the time period from 1990 to 2015. On the other hand, as an index for economic growth, the GDP per capita is considered for the same time period. Both variables refer to the country of Bulgaria. The data employed were derived by the FAOSTAT database. The variables involve the carbon emissions equivalent in thousands of tonnes derived by deforestation for the 1990–2015 time period and per capita income as an index for economic growth.

The variables employed in the study are illustrated in the following Figures 1 and 2, respectively. Figure 1 depicts the evolution of GDP per capita for the time period between 1990–2015. Specifically, there is an evident sharply increasing trend over the period from 2004 to 2005, whereas in the last decade, the GDP per capita has been doubled, and has been stable ever since.

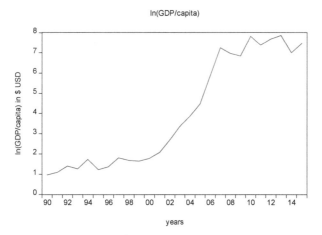

**Figure 1.** ln(GDP/capita) for Bulgaria. GDP: gross domestic product.

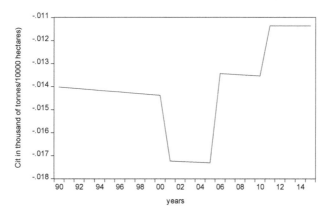

**Figure 2.** Evolution of carbon emissions generated by deforestation (1990–2015) for Bulgaria (ln(Cit).

Figure 2 shows the per capita GDP (in USD), with an upward trend and stability after the year 2010.
Figure 2 shows the evolution of carbon emissions equivalent (in ln), although the pattern of its evolution cannot be defined with accuracy.

## 3.2. ARDL Methodology

The methodology employed is the ARDL bounds cointegration technique. The following ARDL model is used for the scope of this study:

$$\Delta LnC_t = \alpha_0 + \alpha_1 T + \sum_{i=1}^{p-1} \alpha_{1i}\Delta lnC_{t-i} + \sum_{i=0}^{p-1} \alpha_{2i}\Delta lnGDP_{t-i} + \sum_{i=0}^{p-1} \alpha_{3i}\Delta lnGDP^2_{t-i} \\ + \sum_{i=0}^{p-1} \alpha_{4i}\Delta lnGDP^3_{t-i} + \phi_1 lnC_{t-1} + \phi_2 lnGDP_{t-1} + \phi_3 lnGDP^2_{t-1} + \epsilon_t \quad (1)$$

where $C_t$ denotes carbon emissions generated by deforestation, GDP denotes gross added value per capita, T denotes the time trend, $\alpha_0$ the constant, $\Phi_{1,2,3}$ denotes the long-run coefficients, $\alpha_{1i}$, $\alpha_{2i}$, $\alpha_{3i}$, and $\alpha_{4i}$ denote the short run parameters, $\Delta$ denotes the first difference operator of the variable employed, and P denotes the lags determined by the employed lag length optimization criteria such as Akaike Information Criterion (AIC) and Schwarz Bayesian Criterion (SBC). The general model provided above is supplied under the condition that it suffers from stability problems; a modified model might be employed, since the time trend T or the constant coefficient should be omitted.

Equation (2) includes the long-term and the short-term parameters. Implicitly, $\varphi_1$, $\varphi_2$, $\varphi_3$, and $\varphi_4$ denote the long-term parameters, and therefore reject the null hypothesis; $\varphi_1 = \varphi_2 = \varphi_3 = \varphi_4 = 0$ (no cointegration) against the alternate (that is $\varphi_1$, $\varphi_2$, $\varphi_3 \neq 0$) confirms that the variables are cointegrated. The cointegration test is based on the computed F-statistic, being compared to the critical values provided for small size data by [58]. The potential results of this test are synopsized in Table 1.

Table 1. Results of F statistic test.

| Condition | Result |
|---|---|
| Fstatistic > UCB | Cointegration |
| Fstatistic ≤ LCB | No cointegration |
| LCB ≤ Fstatistic ≤ UCB | Uncertain (Result depends on the lagged error correction term for a long-run relationship) |

One of the advantages of the specific methodology involves its efficiency when implemented to variables that have different orders of integration, but this order should be less than two [59]. Therefore, prior to the estimation model and testing the existence of cointegration, it is a prerequisite to explore the order of integration for the variables used in the model. Implicitly, the authors examined the existence of a unit root in the behavior of GDP per capita and carbon emissions equivalent generated by deforestation. The unit root tests employed are the DF-GLS test and the augmented Dickey Fuller (ADF) with break point test. The major advantage of the last-mentioned test involves the simultaneous detection of potential structural breaks along with stationarity for the time series studied.

The break unit root test involves a two-step procedure. In the first step, the authors detrended the series with the appropriate trend and break variables, and also with the assistance of ordinary least squares (OLS), and in the second step, the existence of a unit root with the assistance of a modified Dickey–Fuller regression is implemented to the detrended series [60]. The initial model may involve non-trending or trending data with an intercept break or trend break. The results refer either to the trend or to the break specification (trend, intercept, or both).

Misspecifications are a potential problem of improper unit root test selection. For that reason and for the selection of an optimal method and model of the unit root test, this study used the one proposed by Shrestha and Chowdhury [61].

Having confirmed that the variables employed are not I(2) allowed the authors to implement the ARDL approach. The slightly modified Johansen cointegration technique—namely, the ARDL model—is free from residual serial correlation and endogeneity problems [62]. A number of advantages can be mentioned—including, among others, that the lags employed in the model are selected in a general-to-specific modeling framework [61–63]. Furthermore, the dynamic error correction model (ECM) through a simple linear transformation [64] integrates the short-run dynamics with the long-run equilibrium. Another serious advantage is related to the lack of problems resulting from non-stationary time-series data [63].

Prior to the model estimation and according to the methodology suggested by Pesaran et al. [65], the authors chose the lags for optimum model. The selection of the best model was based on the lowest prediction error. The next step confirmed the existence of a long-run relationship with the assistance of the bounds test and the error correction model.

The long-run relationship to be surveyed is the following for the ARDL (p,q,r) model:

$$\ln C_{it} = \lambda_1 \ln GDP + \lambda_2 \ln GDP^2 + \lambda_3 \ln GDP^3 \, u_t + \mu + \lambda 3 D_t + \delta_t \quad (2)$$

where $\ln C_{it}$ denotes the carbon emissions generated by deforestation for the country to be surveyed, $\ln GDP$ denotes the gross domestic product (GDP) per capita trend in logarithmic form, $\ln GDP^2$ denotes the quadratic form of the GDP per capita, $\ln GDP^3$ denotes the cubic form of the GDP per capita, and $D_t$ is a dummy variable that captures a single structural break. This structural break may be detected in the relationship among the variables studied for Bulgaria, in case that is statistically significant. In case the parameters $\lambda_1$ and $\lambda_2$ are found to be positive and negative respectively as well as statistically significant, the EKC is validated.

In the last step of the analysis, a sensitivity test was performed, through which parameter stability and goodness of fit is tested, which includes the cumulative sum of squares of recursive residuals (CUSUM test). The parameter stability is ensured in case the graphs mentioned above lie within the bounds [65,66].

The error correction model is provided by Equation (3):

$$(1-L) \begin{bmatrix} \ln C_t \\ \ln(GDP)_t \\ \ln(GDP^2)_t \\ \ln(GDP^3)_t \end{bmatrix} = \begin{bmatrix} \varphi_1 \\ \varphi_2 \\ \varphi_3 \\ \varphi_4 \end{bmatrix} + \sum_{i=1}^{p}(1-L)\begin{bmatrix} a_{11i} & a_{12i} & a_{13i} & a_{14i} \\ b_{21i} & b_{22i} & b_{23i} & b_{24i} \\ \delta_{31i} & \delta_{32i} & \delta_{33i} & \delta_{34i} \\ \varepsilon_{41i} & \varepsilon_{42i} & \varepsilon_{43i} & \varepsilon_{44i} \end{bmatrix} + \begin{bmatrix} \beta \\ \zeta \\ \gamma \\ \theta \end{bmatrix} ECM_{t-1} + \begin{bmatrix} \eta_{1i} \\ \eta_{2i} \\ \eta_{3i} \\ \eta_{4i} \end{bmatrix} \quad (3)$$

where (1-L) denotes the lag operator and $ECM_{t-1}$ denotes the lagged error correction term generated by the cointegrating equation, while the η terms are white noises. Short-run causality is confirmed in case the F statistic for the parameters of first differences is statistically significant, whereas the long-run causality is evident in case the lag of the error correction term with the assistance of t–statistics is reported as statistically significant.

## 4. Results

The results of the methodology are provided in the following Tables 2–7. In the first step of the analysis, two different unit roots tests were implemented, namely the DF-GLS unit root test and the DF break unit root test. The results of those tests are provided in Tables 2 and 3, respectively.

Table 2. Elliott–Rothenberg–Stock DF-GLS test.

|  | t–Statistic | 5% Critical Value |
|---|---|---|
| lnCit | −1.11 | −1.955 |
| lnGDP | −1.148 | −1.955 |
| D(Cit) | −4.85 *** | −1.955 |
| D(GDP) | −4.39 *** | −1.955 |

*** Rejection of null hypothesis in 5% level of significance.

**Table 3.** Break unit root test; minimize Dickey–Fuller t-statistic.

|          | t-Statistic | p-Value | Structural Break |
|----------|-------------|---------|------------------|
| lnCit    | −2.964      | 0.71    | 2005             |
| lnGDP    | −2.978      | 0.697   | 2002             |
| D(lnCit) | −6.927 ***  | 0.01    | 2006             |
| D(GDP)   | −4.803 ***  | 0.012   | 2006             |

*** Rejection of null hypothesis in 5% level of significance; Critical value for 5% level of significance: −4.443649.

**Table 4.** F-test estimation result.

| F-Bounds Test  |       | Null Hypothesis: No Levels Relationship | | |
|----------------|-------|---------|------|------|
| Test Statistic | Value | Signif. | I(0) | I(1) |
| F-statistic    | 7.39  | 10%     | 2.37 | 3.2  |
| k              |       | 5%      | 2.79 | 3.67 |
|                |       | 2.5%    | 3.15 | 4.08 |
|                |       | 1%      | 3.65 | 4.66 |

*** Rejection of null hypothesis in 5% level of significance.

**Table 5.** Long-run form.

| Variable    | Coefficient | Std. Error | t-Statistic | Prob.  |
|-------------|-------------|------------|-------------|--------|
| lnGDP       | −13.39 ***  | 0.456      | −29.31      | 0.0000 |
| lnGDP$^2$   | −0.143 ***  | 0.018      | −7.620      | 0.0000 |
| lnGDP$^3$   | 2.65 ***    | 0.190      | 13.96       | 0.0000 |

*** Rejection of null hypothesis in 5% level of significance.

**Table 6.** Error correction model.

| Variable        | Coefficient | Std. Error | t-Statistic | Prob.  |
|-----------------|-------------|------------|-------------|--------|
| D(GDP$^3$) ***  | −0.176      | 0.025144   | −6.452      | 0.0000 |
| D(GDP(-1))      | −0.0035     | 0.002349   | −1.487      | 0.23   |
| D(GDP$^2$) ***  | 2.496       | 0.356405   | 6.583       | 0.00   |
| CointEq(-1) *** | −0.607      | 0.104571   | −5.899      | 0.00   |
| R-squared       | 0.7156      |            |             |        |

*** Rejection of null hypothesis in 5% level of significance.

**Table 7.** Autocorrelation and heteroscedasticity test of the model's residuals.

|                                    | F-Statistic | p-Value |
|------------------------------------|-------------|---------|
| Breusch–Godfrey autocorrelation test | 0.28      | 0.75    |
| ARCH heteroscedasticity test       | 1.118       | 0.268   |

According to these findings, the first unit root test confirmed that the variables used in the model—that is, deforestation as a proxy for environmental degradation and economic growth, respectively—are I(1), which is non-stationary in levels and stationary in first differences.

The second unit root test employed, which takes into consideration the existence of structural breaks, does also confirm that the variables are I(1), as evident in Table 3. Furthermore, this test provides the potential structural breaks of the time-series studied. Based on these, potential explanations for the structural breaks were identified. The year 2006 is a significant hallmark, since it coincides with the end of the privatization of the state-owned firms. This is significant for GDP per capita and deforestation due to a reduction in the foreign direct investments (FDI). In addition, regarding the year 2002, the mechanisms of coordination and management for the implementation of the strategy on structural funds have been refined. Finally, the year 2005 corresponds to when the Kyoto protocol was entered into force, specifically on 16 February 2005, which may adequately interpret the behavior of the carbon emissions equivalent generated by deforestation for the case of Bulgaria.

Having empirically confirmed that the time series studied are not I(2), the ARDL methodology was well deployed. The ARDL model selected based on the Akaike criterion, was ARDL (1,0,2,1). In the next step, the estimated F test suggests that the null hypothesis according to which no level relationships exist cannot be accepted, a result implying the existence of a long-run relationship among the variables studied, as evident in Table 4.

The next step in the analysis involves the estimation of the cointegrating relationship (long-run relationship) model based on the ARDL bounds test and is provided in Table 5.

The estimated error correction term that describes the speed of convergence to the steady state is the following: Cit − (−10.4642*GDP_CAPITA + 2.0064*(GDP)$^2$ − 0.1017*(GDP)$^3$ − 3.0495). The coefficients of the long-run relation are found to be statistically significant (for 10% level of significance), while the signs of the coefficients are as follows; $\lambda_1 < 0$, $\lambda_2 > 0$, and $\lambda_3 < 0$. Therefore, the signs validate the reversed N Kuznets curve pattern. This pattern implies that emissions would begin to rise again once a second income turning point is passed.

The estimation of the error correction model is provided in Table 6. The negative coefficients of $ECT_{t-1}$ are corroborating the short-term relationship in the model. The coefficient of the $ECT_{t-1}$ is indicative of the speed of convergence from short-term disequilibrium to the long-term equilibrium in the approximately 15.5 months in the linear ARDL model. The negative (positive) coefficients of $DGDP_{t-1}$ ($DGDP_{t-12}$) do not confirm the existence of the EKC hypothesis (inverted U pattern) with a one-year lag in the model. Subsequently, the inversed N Kuznets curve is also fully validated for the case of Bulgaria in the short run.

In the short term, it is shown that the statistical significance of the cubic form implies the validity of the N Kuznets curve pattern (not inverted N Kuznets, contrary to the long term). Regarding the diagnostic tests as observed in Table 7, we conducted the Breusch Godfrey autocorrelation test and ARCH heteroscedasticity of the estimated model residuals.

The last step in the analysis involves the study of the parameter's stability with the Cumulative d CUSUM square (CUSUMsq) tests, the results of which are illustrated in Figure 3. Specifically, in Figure 3, the particular plot lies within the critical bounds at a 5% significance level, which indicates that the estimated model is stable in the research period.

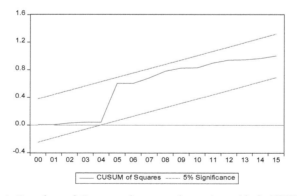

**Figure 3.** Illustration of cumulative sum of squares of recursive residuals (CUSUM) of squares stability test.

Furthermore, the parameter stability CUSUM of squares and the diagnostic tests provided in Table 7 confirm the robustness of the linear ARDL estimates test.

## 5. Conclusions

Among EU members, Bulgaria was selected as one of the newly entrant economies, having a high growth rate and specific condition in the deforestation process. This study investigated the validity of EKC in alignment with the ARDL approach of carbon emissions generated due to deforestation (in thousands of tonnes per 1000 ha of deforested ex forest land) as an index of environmental degradation, and the per capita GDP (in thousand of USD dollars) as an index for income. According to these findings, a reverted N Kuznets curve pattern was validated, which was a result implying two different income thresholds for change in the behavior of carbon emissions. This result was validated not only for the long-term period, but also for the N pattern relationship for the short term. Having confirmed the validity of the opposite N-shaped EKC implies that the economic growth initially will improve environmental quality to a certain income level, where the relationship will be positive before it once again becomes negative. This finding is not only interesting, but also challenging to interpret. Possibly, it could be a consequence of the initial environmentally-friendly attitude of the Bulgarian economy and its limitations toward deforestation as well as the efforts aiming to afforestation, compensating for the increased emissions caused by the scale effect. The results were based on carbon emissions generated by deforestation, and these may well efficiently and indirectly describe the forest land situation in Bulgaria, and contribute effectively to the design and implementation of environmental policies that are capable of eliminating the deforestation problem in Bulgaria.

The process of afforestation could provide a solution to the problem of deforestation—namely, a reduction in GHG emissions and an increased absorption of environmental pollutants. In addition, policies and measures design and implementation should promote the efficiency of human activities in order to ensure that the economic losses attributed to the limited exploitation of forest resources are limited.

The state's efforts for afforestation (a plausible explanation for the findings) were outperformed by the illegal logging, and the low wages of the civil servants in the forest service may also lead to a higher quantity of carbon emissions generated by deforestation. Therefore, initiatives should be taken for the counteraction of these behaviors that can be achieved by seminars, in order for residents to be better informed, and a provision of motivation for them to pursue a limitation in deforestation. This motivation could include, among others, economic incentives for the conversion of forestland to agricultural uses such as taxes or subsidies. Furthermore, clearly defined and enforced property rights to timberlands could also provide an effective solution to the limitation of the problem.

To synopsize the novel institutional measures that should be taken, we aim to limit the degradation of forest resources, which in turn will bring about a decrease in the carbon emissions generated by deforestation. Furthermore, the smooth forest land use with alternative methodologies may well lead to quality environmental improvement and the prosperity of rural and urban surrounding areas.

The specific conditions that dominate in Bulgaria despite the country complying with the agri-environmental measures adopted by the EU may also interpret the opposite N pattern of environmental performance–GDP per capita relationship, necessitating a more insightful study regarding the formation of the variables and the evolution of this relationship. For that reason, the findings of this study contradict the findings of Zambrano-Monserrate et al. [67] that confirmed the validity of EKC for other five European countries with the same methodology for a longer time horizon. Based on the aforementioned results, the environmental policies mainly for the case of Bulgaria should be directed to the expansion of forest land, since forestry production and agricultural exports may increase jointly, allowing a significant progress for environmental protection inside the continent.

The study also unveiled a remarkable conclusion regarding the shape of the EKC, thus suggesting research intensification in order for the pollution–income relationship to be identified. The relationship may well be studied in alignment with an alternative methodology while considering other possible shapes than those already examined and expected in other EKC core studies. It is important to further investigate the relationship between income and environmental degradation in order to combat climate change and reach sustainable economic development.

Conclusively, regarding suggestions for the future research, the implementation of a different nonlinear ARDL methodology on available data could provide more concise and accurate results, as well as the implementation of panel data analysis to support researchers with more general results regarding the agro-environmental EU policy for more countries.

**Author Contributions:** S.T. organized the study, conducted the design of the study, and analyzed the data. E.Z. deployed the analysis, the interpretation of results, as well as the revision of the study. E.Z. and G.K. modified and smoothed the written English. G.K. and G.A. deployed the literature review. S.T., E.Z., G.K., and G.A. formulated and finalized the conclusions of the study.

**Funding:** This research received no external funding.

**Conflicts of Interest:** The authors declare no conflicts of interest.

## References

1. Bonan, G.B. Forests and climate change: Forcings, feedbacks, and the climate benefits of forests. *Science* **2008**, *320*, 1444–1449. [CrossRef] [PubMed]
2. Skoutaras, D. Tropical Forests: Problems and Their Possibility of Confrontation. Master's Thesis, Department of Planning and Regional Development, School of Engineering, University of Thessaly, Volos, Greece, 2010; 107p.
3. Stern, N.H. *The Economics of Climate Change: The Stern Review*; Cambridge University Press: Cambridge, UK, 2007.
4. Van der Werf, G.R.; Morton, D.C.; DeFries, R.S.; Olivier, J.G.J.; Kasibhatla, P.S.; Jackson, R.B.; Collatz, G.J.; Randerson, J.T. $CO_2$ Emissions from Forest Loss. *Nat. Geosci.* **2009**, *2*, 737–738. [CrossRef]
5. Thompson, I.; Mackey, B.; McNulty, S.; Mosseler, A. *Forest Resilience, Biodiversity, and Climate Change: A Synthesis of the Biodiversity/Resilience/Stability Relationship in Forest Ecosystems*; Technical Series no. 43; Secretariat of the Convention on Biological Diversity: Montreal, QC, Canada, 2009; 67p.
6. Fady, B.; Cottrell, J.; Ackzell, L.; Alía, R.; Muys, B.; Prada, A.; González-Martínez, S.C. Forests and global change: What can genetics contribute to the major forest management and policy challenges of the twenty-first century? *Reg. Environ. Chang.* **2016**, *16*, 927–939. [CrossRef]
7. Pretzsch, H.; Biber, P.; Schutze, G.; Uhl, E.; Rötzer, T. Forest stand growth dynamics in Central Europe have accelerated since 1870. *Nat. Commun.* **2014**, *5*, 1–10. [CrossRef] [PubMed]
8. Main-Knorn, M.; Hostert, P.; Kozak, J.; Kuemmerle, T. How pollution legacies and land use histories shape post-communist forest cover trends in the western Carpathians. *For. Ecol. Manag.* **2009**, *258*, 60–70. [CrossRef]
9. Chew, S.C. *World Ecological Degradation: Accumulation, Urbanization, and Deforestation, 3000 BC–AD 2000*; AltaMira Press: Walnut Creek, CA, USA, 2001; 228p.
10. Rudel, T.K. The national determinants of deforestation in sub-Saharan Africa. *Philos. Trans. R. Soc. Lond. B Biol. Sci.* **2013**, *368*, 20120405. [CrossRef] [PubMed]
11. Lofdahl, C. *Environmental Impacts of Globalization and Trade*; MIT Press: Cambridge, MA, USA, 2002.
12. World Bank. World Development Indicators. 2005. Available online: http://www.worldbank.org (accessed on 20 May 2019).
13. Kindermann, G.; Obersteiner, M.; Sohngen, B.; Sathaye, J.; Andrasko, K.; Rametsteiner, E.; Beach, R. Global cost estimates of reducing carbon emissions through avoided deforestation. *PNAS* **2008**, *105*, 10302–10307. [CrossRef] [PubMed]
14. Bala, G.; Caldeira, K.; Wickett, M.; Phillips, T.J.; Lobell, D.B.; Delire, C.; Mirin, A. Combined climate and carbon-cycle effects of large-scale deforestation. *PNAS* **2007**, *104*, 6550–6555. [CrossRef] [PubMed]
15. FAO. *State of the World's Forests 2016—Forests and Agriculture: Land-Use Challenges and Opportunities*; FAO: Rome, Italy, 2016; 126p.
16. Dimitrova, A.; Buzogány, A. Post-accession policy-making in Bulgaria and Romania: Can non-state actors use EU rules to promote better governance? *J. Common Market Stud.* **2014**, *52*, 139–156. [CrossRef]
17. FAO. *Global Forest Resources Assessment 2015*; FAO: Rome, Italy, 2015; 253p.

18. Allen, J.C.; Barnes, D.F. The causes of deforestation in developing countries. *Ann. Assoc. Am. Geogr.* **1985**, *75*, 163–184. [CrossRef]
19. Indarto, J.; Mutaqin, D.J. An overview of theoretical and empirical studies on deforestation. *J. Int. Dev. Coop.* **2016**, *22*, 107–120.
20. Kaika, D.; Zervas, E. The Environmental Kuznets Curve (EKC) theory—Part A: Concept, causes and the $CO_2$ emissions case. *Energy Policy* **2013**, *62*, 1392–1402. [CrossRef]
21. Stern, D.I. The rise and fall of the Environmental Kuznets Curve. *World Dev.* **2004**, *32*, 1419–1439. [CrossRef]
22. Mahmood, H.; Maalel, N.; Zarrad, O. Trade Openness and $CO_2$ Emissions: Evidence from Tunisia. *Sustainability* **2019**, *11*, 3295. [CrossRef]
23. Soytas, U.; Sari, R.; Bradley, T.; Ewing, B.T. Energy consumption, income, and carbon emissions in the United States. *Ecol. Econ.* **2007**, *62*, 482–489. [CrossRef]
24. Grossman, G.M.; Krueger, A.B. *Environmental Impact of a North American Free Trade Agreement*; Working Paper 3914; National Bureau of Economic Research: Cambridge, MA, USA, 1991.
25. Shahbaz, M.; Sinha, A. *Environmental Kuznets Curve for $CO_2$ Emission: A Literature Survey*; MPRA Paper No. 86281; University Library of Munich: Munich, Germany, 2018; 83p.
26. Apergis, N.; Payne, J.E. Energy consumption and economic growth in Central America: Evidence from a panel cointegration and error correction model. *Energy Econ.* **2009**, *31*, 211–216. [CrossRef]
27. Aruga, K. Investigating the Energy-Environmental Kuznets Curve Hypothesis for the Asia-Pacific Region. *Sustainability* **2019**, *11*, 2395. [CrossRef]
28. Ang, J. $CO_2$ emissions, energy consumption, and output in France. *Energy Policy* **2007**, *35*, 4772–4778. [CrossRef]
29. Ozatac, N.; Gokmenoglu, K.K.; Taspinar, N. Testing the EKC hypothesis by considering trade openness, urbanization, and financial development: The case of Turkey. *Environ. Sci. Pollut. Res.* **2017**, *24*, 16690–16701. [CrossRef]
30. Akbostancı, E.; Türüt-AsIk, E.; Ipek Tunc, G.I. The relationship between income and environment in Turkey: Is there an environmental Kuznets curve? *Energy Policy* **2009**, *37*, 861–867. [CrossRef]
31. Liu, X.; Zhang, S.; Bae, J. The impact of renewable energy and agriculture on carbon dioxide emissions: Investigating the environmental Kuznets curve in four selected ASEAN countries. *J. Clean. Prod.* **2017**, *164*, 1239–1247. [CrossRef]
32. Cropper, M.; Griffiths, C. The Interaction of Population Growth and Environmental Quality. *Am. Econ. Rev.* **1994**, *84*, 250–254.
33. Joshi, P.; Beck, K. Environmental kuznets curve for deforestation: Evidence using GMM estimation for OECD and non-OECD regions. *iForest Biogeosci. For.* **2017**, *10*, 196–203. [CrossRef]
34. Choumert, J.; Combes Motel, P.; Dakpo, H.K. Is the Environmental Kuznets Curve for deforestation a threatened theory? A meta-analysis of the literature. *Ecol. Econ.* **2013**, *90*, 19–28. [CrossRef]
35. Lynch, M.J. A Marxian Interpretation of the Environmental Kuznets Curve: Global Capitalism and the Rise and Fall (and Rise) of Pollution. *Capital. Nat. Soc.* **2016**, *27*, 77–95. [CrossRef]
36. Esmaeili, A.; Nasrnia, F. Deforestation and the Environmental Kuznets Curve in Iran. *Small Scale For.* **2014**, *13*, 397–406. [CrossRef]
37. Ogundari, K.; Ademuwagun, A.A.; Ajao, O.A. Revisiting environmental kuznets curve in Sub-Sahara Africa: Evidence from deforestation and all GHG emissions from agriculture. *Int. J. Soc. Econ.* **2017**, *44*, 222–231. [CrossRef]
38. Zafeiriou, E.; Sofios, S.; Partalidou, X. Environmental Kuznets curve for EU agriculture: Empirical evidence from new entrant EU countries. *Environ. Sci. Pollut. Res.* **2017**, *24*, 15510–15520. [CrossRef]
39. Barbier, E.B.; Burgess, J.C. The Economics of Tropical Deforestation. *J. Econ. Surv.* **2001**, *15*, 413–433. [CrossRef]
40. Ehrhardt–Martinez, K.; Crenshaw, E.M.; Jenkins, J.C. Deforestation and the Environmental Kuznets Curve: A Cross-National Investigation of Intervening Mechanisms. *Soc. Sci. Q.* **2003**, *83*, 226–243. [CrossRef]
41. Lantz, V. Is there an Environmental Kuznets Curve for Clear-cutting in Canadian Forests? *J. For. Econ.* **2002**, *8*, 199–212.
42. Chiu, Y.B. Deforestation and the environmental kuznets curve in developing countries: A panel smooth transition regression approach. *Can. J. Agric. Econ./Rev. Can. D'agroecon.* **2012**, *60*, 177–194. [CrossRef]

43. Farhani, S.; Mrizak, S.; Chaibi, A.; Rault, C. The environmental Kuznets curve and sustainability: A panel data analysis. *Energy Policy.* **2014**, *71*, 189–198. [CrossRef]
44. Parajuli, R.; Omkar Joshi, O.; Tek Maraseni, T. Incorporating Forests, Agriculture, and Energy Consumption in the Framework of the Environmental Kuznets Curve: A Dynamic Panel Data Approach. *Sustainability* **2019**, *11*, 2688. [CrossRef]
45. Zafeiriou, E.; Azam, M. $CO_2$ emissions and economic performance in EU agriculture: Some evidence from Mediterranean countries. *Ecol. Indic.* **2017**, *81*, 104–114. [CrossRef]
46. Zafeiriou, E.; Katrakilidis, C.; Pegiou, C. Consumer Confidence on Heating Oil Prices: An Empirical Study of their Relationship for European Union in a Nonlinear Framework. *Eur. Res. Stud.* **2019**, *22*, 68–95.
47. Grossman, G.; Krueger, A. Economic environment and the economic growth. *Q. J. Econ.* **1995**, *110*, 353–377. [CrossRef]
48. Tiwari, A.K.; Shahbaz, M.; Hye, Q.M.A. The Environmental Kuznets curve and the role of coal consumption in India: Cointegration and causality analysis in an open economy. *Renew. Sustain. Energy Rev.* **2013**, *18*, 519–527. [CrossRef]
49. Onafowora, O.A.; Owoye, O. Bounds testing approach to analysis of the environment Kuznets Curve hypothesis. *Energy Econ.* **2014**, *44*, 47–62. [CrossRef]
50. Jebli, M.B.; Youssef, S.B.; Ozturk, I. Testing Environmental Kuznets Curve hypothesis: The role of renewable and non-renewable energy consumption and trade in OECD Countries. *Ecol. Indic.* **2016**, *60*, 824–831. [CrossRef]
51. Zambrano-Monserrate, M.A.; Valverde-Bajana, I.; Aguilar-Bohorquez, J.; Mendoza-Jimenez, M.J. Relationship between economic growth and environmental degradation: Is there evidence of an environmental Kuznets curve for Brazil? *Int. J. Energy Econ. Policy.* **2016**, *6*, 208–216.
52. Ercilla–Montserrat, M.; Muñoz, P.; Montero, J.-I.; Gabarrell, X.; Villalba, G. A study on air quality and heavy metals content of urban food produced in a Mediterranean city (Barcelona). *J. Clean. Prod.* **2018**, *195*, 385–395. [CrossRef]
53. Kaygusuz, K. Energy and environmental issues relating to greenhouse gas emissions for sustainable development in Turkey. *Renew. Sustain. Energy Rev.* **2009**, *13*, 253–270. [CrossRef]
54. Esteve, V.; Tamarit, C. Is there an environmental Kuznets curve for Spain? Fresh evidence from old data. *Econ. Model.* **2012**, *29*, 2696–2703. [CrossRef]
55. Ozturk, I.; Acaravci, A. $CO_2$ emissions, energy consumption and economic growth in Turkey. *Renew. Sustain. Energy Rev.* **2010**, *14*, 3220–3225. [CrossRef]
56. Halicioglu, F. An econometric study of $CO_2$ emissions, energy consumption, income and foreign trade in Turkey. *Energy Policy* **2009**, *37*, 1156–1164. [CrossRef]
57. Soytas, U.; Sari, R. Energy consumption and GDP: Causality relationship in G-7 countries and emerging markets. *Energy Econ.* **2003**, *25*, 33–37. [CrossRef]
58. Narayan, P.K. The saving and investment Nexus for China: Evidence from cointegration tests. *Appl. Econ.* **2005**, *37*, 1979–1990. [CrossRef]
59. Pesaran, M.H.; Pesaran, B. *Working with Microfit 4.0: Interactive Econometric Analysis*; Oxford University Press: Oxford, UK, 1997.
60. Vogelsang, T.J.; Perron, P. Additional Tests for a Unit Root Allowing for a Break in the Trend Function at an Unknown Time. *Int. Econ. Rev.* **1998**, *39*, 1073–1100. [CrossRef]
61. Shrestha, M.B.; Chowdhury, K. *ARDL Modelling Approach to Testing the Financial Liberalisation Hypothesis*; Working Paper 05-15; Department of Economics, University of Wollongong: Wollongong, Australia, 2005; 33p.
62. Pesaran, M.H.; Shin, Y. An autoregressive distributed lag modelling approach to cointegration analysis. In *Econometrics and Economic Theory in the 20th Century: The Ragnar Frisch Centennial Symposium*; Strom, S., Ed.; Cambridge University Press: Cambridge, UK, 1999.
63. Laurenceson, J.; Chai, J. *Financial Reform and Economic Development in China*; Edward Elgar: Cheltenham, UK, 2003.
64. Banerjee, A.; Dolado, J.; Galbraith, J.; Hendry, D. *Co-Integration, Error Correction, and the Econometric Analysis of Non-Stationary Data*; Oxford University Press: Oxford, UK, 1993.
65. Pesaran, M.H.; Shin, Y.; Smith, R.J. Bounds testing approaches to the analysis of level relationships. *J. Appl. Econ.* **2001**, *16*, 289–326. [CrossRef]

66. Pesaran, M.; Shin, Y.; Smith, R. Structural analysis of vector error correction models with exogenous I(1) variables. *J. Econ.* **2000**, *97*, 293–343. [CrossRef]
67. Zambrano-Monserrate, M.A.; Carvajal-Lara, C.; Urgilés-Sanchez, R.; Ruano, M.A. Deforestation as an indicator of environmental degradation: Analysis of five European countries. *Ecol. Indic.* **2018**, *90*, 1–8. [CrossRef]

© 2019 by the authors. Licensee MDPI, Basel, Switzerland. This article is an open access article distributed under the terms and conditions of the Creative Commons Attribution (CC BY) license (http://creativecommons.org/licenses/by/4.0/).

Article

# An Investigation of Factors Affecting the Willingness to Invest in Renewables among Environmental Students: A Logistic Regression Approach

Evangelia Karasmanaki *, Spyridon Galatsidas and Georgios Tsantopoulos

Department of Forestry and Management of the Environment and Natural Resources, Democritus University of Thrace, 68200 Orestiada, Greece; sgalatsi@fmenr.duth.gr (S.G.); tsantopo@fmenr.duth.gr (G.T.)
* Correspondence: evagkara2@fmenr.duth.gr; Tel.: +30-6-983-605-600

Received: 16 August 2019; Accepted: 10 September 2019; Published: 13 September 2019

**Abstract:** Renewable energy sources (RES) have gained increased popularity across the world mainly due to their ability to contribute to environmental protection through the generation of infinite 'clean' energy. To achieve a greater diffusion of renewables, however, small-scale investments implemented by individuals are critically important. In contrast to citizens whose attitudes have been consistently explored by research, there is little evidence on the attitudes towards investments among environmental students who will occupy positions of responsibility and play key roles in the environmental sector in the future. Hence, the purpose of the present study is to identify the most important factors that affect environmental students' willingness to invest in renewable energy (RE) by developing a logistic regression model. According to our analysis, the participants in their majority expressed their willingness to invest, while environmental values, the low risk and profitability of renewable investments, as well as preferences for certain energy types were significant factors determining this willingness. However, willingness to invest was irrespective of the current taxation and subsidies, suggesting that significant improvements are required in these areas. The present study could be particularly useful for policymakers since the necessary steps to create favorable investment environments in Greece and elsewhere are highlighted.

**Keywords:** willingness-to-invest; energy investments; renewable energy sources (RES); perceptions; attitudes; department of forestry; environmental science; university students

## 1. Introduction

For a fourth consecutive year, the worldwide net capacity additions for renewable energy significantly exceeded those for fossil fuels and nuclear power, while renewables accounted for more than one-third of the total global installed power capacity in 2018. During the same year, a total of USD 36.3 billion was invested in small-scale distributed capacity (i.e., investments in solar photovoltaics systems smaller than 1 MW), a 15% decrease compared to 2017 [1].

Regarding Greece, the development of renewable energies shows a definite upturn [2] while the Renewables 2019 Global Status Report placed the country among the nine countries in the world which present more than 20% electricity production from renewables [1]. According to the latest reports [3], in Greece, the share of renewable sources accounted for 31% of electricity production in 2016 with the main renewable sources being solar, wind, and hydropower. More analytically, solar power production witnessed a striking growth since it rose from 0.16 TWh in 2010 to 3.9 TWh in 2016. Likewise, wind energy production increased from almost nonexistent levels in the end of the 1990s to 5.1 TWh in 2016, which corresponds to 10.5% of the total electricity production. At the same time, hydropower has been experiencing a steady increase in its share of renewable electricity, but with significant annual fluctuations—reaching 5.5 TWh in 2016 which accounts for 11.4% of the total production. During the same year, the share of electricity from bioenergy covered less than 1% of the total production [3]. At the same time, the country is renowned for its impressive implementation of solar installations and Greece is one of the pioneers in terms of solar systems' use, with the first systems being installed in the late 1970s, whereas in 2016, the country had already achieved one of the highest shares of solar photovoltaics in the total primary energy supply among IEA countries [3].

The impressive share increases mentioned above were driven mainly by supporting policies and in specific by a series of feed-in-laws, namely Law No 3468/2006, Law No 3734/2009, and Law No 3851/2010 [4–6]. These laws established fixed prices for renewable electricity, forming an attractive investment environment. Law No 3851/2010 is still in effect today and was initially established to address the numerous pending applications for the approval of photovoltaic system applications which had brought the domestic photovoltaic market close to collapse. The same law also foresaw the simplification and acceleration of administrative processes which concerned the permits of installations. In addition to the feed-in laws and the simplification of procedures, various programs and mechanisms have been introduced to aid the development of renewable energies in Greece. For instance, the Financial Program "Save Energy at Home II" provides financial support for upgrading the energy of residences. Beneficiaries, who are selected based on the residence's initial energy category and their income, are offered support through fractional subsidization combined with favoring bank loans from contracting bank institutions. The actions following the provision of finance enhance the energy of the residence with the focus being placed on shell improvements and upgrades in heating/cooling systems as well as warm water usage [7]. Simultaneously, the country is designing competitive auctions for larger wind and solar installations as well as market-based premiums to prevent high-cost overruns [3]. With regard to these new policies and measures, the renewable sector can be expected to experience further development and attract considerable small- and large-scale investments which will in turn help the country to further increase the share of its renewable electricity production.

Beside large investments implemented by the state and private investors, small-scale investments made by citizens can contribute significantly to the achievement of the targets and the rapid diffusion of renewables [8]. To increase small-scale investments, however, citizens need to be adequately willing to invest their savings or part of their income in renewable energies. In Greece, the public responded positively to incentives for photovoltaic systems (enacted with Law 3468/2006) and laws forming an attractive investment and licensing climate. Moreover, the applications were so numerous that the licensing for specific photovoltaic categories had to be suspended [6]. This reflected the increased interest of householders to invest in their own microgeneration system when the conditions in terms of incentives and legislation were deemed favorable. In addition, it is interesting to observe that citizens' positive response to investments in photovoltaics was recorded during the time of economic crisis.

In a broader perspective, the public needs to be positively inclined towards investments in environmentally friendly energy technologies to proceed with the investment. Acknowledging the effect of personal attitudes on decision-making regarding investments, a growing body of research has been examining attitudes indicating that attitudes comprise the key to predict investment behaviors [9,10].

In the relevant literature, most previous studies have typically focused on the willingness of individuals to pay for renewable energy [11–19], while a substantial number of studies has explored individuals' willingness to invest in renewable energy [9,20–24]. As observed, no study has so far examined explicitly young respondent groups such as undergraduate students whose study field is closely related to the environment. Nevertheless, it is quite relevant to examine the attitudes of students majoring in environmental sciences not only because these students have an adequate understanding of energy topics, but mainly because in their future careers they will be required to take a stand towards renewable energy and their decisions will affect the deployment of renewable energy. Hence, the primary aim of the present study is to identify the factors that affect the willingness to invest in renewables among students majoring in the Department of Forestry and Management of the Environment and Natural Resources at the Democritus University of Thrace. Moreover, the study also builds participants' profiles by collecting data relating to their sociodemographic characteristics and their views on energy-related topics.

The present study and its findings could be particularly useful to policymakers, developers, and marketers in their efforts to create favorable investment environment in Greece and elsewhere. In addition, our results reveal new insights into individuals' investment willingness, thereby contributing to the relevant literature strand and enabling researchers to build on the findings to carry out further studies. Furthermore, the novelty of the present paper is the focus on students' awareness about RES by examining the factors that affect their willingness to invest in renewable energy in their later life through the performance of logistic regression analysis.

The paper is structured as follows. In the next section, the findings of previous relevant studies are presented and discussed. Then, Section 3 describes the area of study as well as the methodology the authors followed to perform the study and to develop the logistic regression model. The results are thoroughly presented in Section 4 and discussed in Section 5. Afterwards, in Section 6, conclusions based on our analysis and discussion are drawn.

## 2. Literature Review

The diffusion of renewable energies is a prerequisite for achieving a low-carbon energy system that can alleviate pressing environmental issues such as climate change [25,26]. Nowadays, citizens are given the option to participate in this endeavor by investing in renewable energy and, in this regard, discovering the attitudes of the public towards investments is an integral part of strategies aimed at increasing investments. Acknowledging the role of the public, a growing body of literature has examined its perceptions and attitudes in an attempt to identify the factors that affect the willingness to invest in renewables. Overall, these studies illustrate that environmental values, confidence in the technology of renewable energy systems, previous experience in investments, age, place of residence, educational level, preference for comfort, social class, and house ownership are factors that can determine the investment willingness of the public [9,10,20,24,27–29].

In terms of environmental values, in countries which are known for their successful implementation of renewable energy, citizens were mainly driven by their environmental values to invest in RES. Indicatively, in a qualitative study on household adoption of small-scale electricity production in Sweden, environmental concerns were what motivated adopters' decisions to install the systems, while the respondents emphasized the need to think about the environment and live 'as ecologically sustainably as possible' [30]. Similar results emerged in another study in the US [31] with the development of a fuzzy logic reference model indicating that for consumers who had already adopted photovoltaics in US, the perceived environmental benefit was positively related to their decision-making. Likewise, in Canada it was observed that environmental values could predict the adoption of sustainable technologies among citizens [27]. The important role of environmental values was also detected in Austria, Italy, and Germany, where relevant studies examining the socio-psychological patterns of RES investments concluded that the desire to make a contribution to environmental protection comprised a key factor for making investments favoring the environment [23,29]. It is also noteworthy that environmental awareness was identified as an important investment factor in countries which present a relatively lower installed RES capacity such as Tunisia, Egypt, Lebanon, Jordan, [28] and Turkey [32]. This indication suggests that environmental awareness is universally influential and to some degree independent of financial restrictions.

Apart from environmental awareness, attitudes towards the financial aspect of investments in renewables are important to discuss. As indicated, the decision to make investments is often based on the individual's evaluation of the expected profitability which if deemed adequate, the investment proceeds [33]. This resonates with Bergek et al. [8] who claimed that investments in renewable energy are made when an opportunity is identified and when the value of exploiting this opportunity is high enough. Building on this, the same research team concluded that the potential value of an opportunity, namely the profit, is a notable motive behind the decision, while accessibility of financial resources affects further investment willingness. At the same time, although incentives had initially paved the way to citizen investment, their long-term effectiveness is often disputed. On one hand, financial incentives have undoubtedly assisted the diffusion of renewable energies across different countries [32,34]. Nevertheless, the positive impacts of financial incentives are often unexpectedly temporary and unstable because individuals tend to revert to their previous behavior once rewards are removed, whereas individuals who are intrinsically motivated to act altruistically respond negatively to extrinsic rewards [35].

In contrast to the positive influence of environmental values and perceived profitability, the lack of trust in the technology of renewable systems can affect adversely an individual's decision to invest. In particular, it was indicated that potential investors' low degree of confidence in the effectiveness of the technology was negatively related to their decision-making [24]. Furthermore, homeowners were less likely to invest when they misunderstood the functioning of the technology [29] or when they considered that the renewable systems were new on the market and therefore disadvantageous to purchase [30]. The lack of confidence can thus to some extent account for the fact that the diffusion of RES is relatively slow and that microgeneration technologies are often described as 'resistant innovations' in many European countries [10].

As with all types of investments, experience is considered an influential factor in the final decision to invest in renewables. However, existing research findings are conflicting about whether previous investment experience causes positive or negative responses to RE investments. In particular, Leete et al. [36] found that investors who were experienced in marine renewable energy were less likely to do so again since they understood the scale, cost unpredictability, and the required time for developing these technologies. Conversely, Masini and Menichetti [24] as well as Ek et al. [33] argued that former experience with investments increases the likelihood to invest in renewable energy. Regardless of these studies, it appears that in the case of photovoltaic panels, which are widely implemented in some European countries, such as Germany and Denmark [37], being able to 'experience' and observe the functioning of the systems on the rooftops of neighboring houses can stimulate the interest of individuals and induce them to invest [27] as was the case with California where the early installation of solar photovoltaic panels in the neighborhood of potential investors rapidly diffused the installation of solar panels [38].

In addition to the above factors, the age of potential investors has been found to affect an individual's inclination to invest; however, age seems to exert an ambiguous influence on investment decisions. On one hand, there is evidence that younger individuals are more open to RES compared to their older counterparts [24]; but, on the other hand, different findings indicated that older people were more likely to be aware of microgeneration technologies and hence more likely to adopt them [10]. That being said, recent findings detected that older respondents were unwilling to invest in renewable technologies probably because they were discouraged by the long payback periods [28]. It can thus be suggested that the influence of age as a factor affecting investments is multifaceted and unclear, thereby calling for further investigation.

Another factor that has been found to be substantially influential is the type of place an individual resides with studies showing that rural dwellers are generally more positively inclined towards renewable investments [39]. For example, city inhabitants in Ireland were unaware of renewable microtechnologies [10], whereas rural residents in Germany were more willing to switch from their current conventional energy situation to a sustainable domestic energy system compared to the city inhabitants who participated in the same study [29]. The same applied to Swedish respondents living in rural areas who regarded investments in micropower plants as a 'logical and practical option' which enabled them to become self-sufficient and energy independent but also resonated with their chosen lifestyle and wish to use the available natural resources [30].

Other variables affecting investment willingness encompass educational level, house ownership, as well as preference for maintaining comfort, status quo, and social class. Of these, high educational level [29,34] and house ownership [23] have been found to have a positive effect on an individual's willingness to invest in renewable energy, while higher educational levels are also linked to increased concerns about the ecological impacts of renewable installations [40]. Meanwhile, in their decision to change their current fossil fuel-based energy system, homeowners can be affected by social norms [9] and their wish to maintain their accustomed comfort and current status quo [29]. In relation to this point, the application of behavioral economics revealed that individuals have the tendency to make social comparisons, follow other people's behavior, and comply with social norms, namely the explicit and implicit rules, guidelines, and behavioral expectations [35]. In this regard, individuals are expected to invest if other people in their close environment have done so and vice versa. As for social class, so far it has been indicated that those belonging to the upper-middle classes are more aware of microgeneration technologies [10] and hence more likely to adopt them.

As observed, the literature on investment willingness relating to renewable energy has mainly focused on analyzing the perceptions and intentions of citizens in different countries. In relation to university students, there is a plethora of relevant studies exploring students' knowledge, perceptions, and attitudes towards renewables in different countries [41–46], whereas the existing studies analyzing undergraduate students' willingness either to pay for or invest in renewable energy are scarce. Indicatively, so far it has been indicated that Chinese undergraduate students were willing to pay additional amounts of money for clean energy or for switching from their current conventional energy systems to renewable energy-based or forest bioenergy-based systems [43]. Likewise, Canadian and Romanian students were positively inclined towards paying for renewable energy-based electricity [47]. Meanwhile, a considerable share of Palestinian university students (60.3%) expressed their willingness to invest in RES projects if such an opportunity emerged, while 71.3% of respondents regarded investments in these projects as successful [48]. It can be seen that only a few studies have, inter alia, examined the investment willingness of university students while no study has performed further analyses to understand the factors that determine their willingness or unwillingness to make renewable investments in the future.

The main conclusion to be drawn from the previous research efforts on attitudes towards investments in renewables, given the aim of this paper, is that overall there exists substantial interest in this special type of investments. With the exception of environmental awareness, which has a profound impact on potential investors, certain factors have emerged as complex and thus further research is needed to understand them. Their investigation should, however, not be neglected because these factors can pave the way for increasing investments. In addition, the understanding of potential investors' mindset can become an effective tool in the hands of policymakers and developers to create a favorable investment environment that will attract more investments among citizens who wish to make profit, but also citizens who look to express their environmental values through investing in environmentally friendly energy production systems.

## 3. Materials and Methods

### 3.1. Area of Study

The Area of Study was the Department of Forestry and Management of the Environment and Natural Resources of the Democritus University of Thrace, which is located in the town of Orestiada in northern Greece. The Department was founded in the academic year 1999–2000 and its current enrollment rate is 105 new students per year. It is worthwhile to note that the Department of Forestry and Management of the Environment and Natural Resources aims at promoting the Sciences of Forestry and the Environment while advancing scientific knowledge in the area of Natural Resources Management. The graduates are highly qualified and capable of conducting research and using advanced technologies for the development, improvement, protection, and management of forests, forest lands, and the natural environment. Another objective of the Department is to contribute to the development of the Science of Forestry through academic training, research, scientific publications, and textbook development. Finally, the Department and its graduates can help improve the management of forests in Greece and protect the country's natural environment. The duration of the undergraduate studies is five years and the curriculum includes multiple courses. Beside the explicitly environmentally- and forestry-related courses, students also attend renewable energy-related courses, such as "Renewable Energy Sources", "Energy and Environment", and "Forest Energy". In addition, there are courses which aim at enhancing students' environmental awareness such as "Environmental Education", "Environmental Communication", "Environmental Policy", and "Forest Policy" [26].

*3.2. Methods*

The findings which are presented in this paper are part of wider research which investigated the awareness levels of the students majoring in the Department of Forestry and Management of the Environment and Natural Resources about renewable energy sources.

For the purposes of the study, a structured questionnaire consisting of 21 closed-ended questions was designed (the full version of the questionnaire can be seen in Karasmanaki and Tsantopoulos [26]). The closed-ended type was considered appropriate since it requires little time and effort to be completed. To ensure that the questionnaire could give coherent and accurate results, it was tested on a limited scale and minor changes were made. Once the final form was ready, the questionnaire was administered to the participants at the beginning of regular class periods with the consent of the professors. The respondents of the sample were undergraduate students of all years of study and, in total, 214 students took part in the study.

The questionnaire consisted of six sections with the first involving a set of questions collecting participants' sociodemographic characteristics. The second section included questions which investigated students' awareness about RES but also their views on various environmentally- and energy-related topics. Then, the third section examined their willingness to invest in renewables, to work in the RES sector and to pay for the development of RES. Afterwards, the fourth section explored respondents' feelings about energy scarcity due to resource depletion and their preferences for various energy types. The fifth section investigated their perceptions of ways and approaches to transition to a low-carbon energy system as well as reasons to adopt RES. Finally, the last section explored their views on environmental responsibility and their preferences for different types of media. It should also be noted that the questions were rated on five-, six-, and ten-point Likert scales, while there were also questions with dichotomous ("yes/no") answers.

Once the collection of the questionnaires was completed, the data were analyzed with the Statistical Package for Social Sciences (SPSS). Initially, descriptive statistics were estimated for all variables. Then, logistic regression was employed to estimate the parameters of a logistic model. According to Hosmer et al. [49] logistic regression is a statistical method for analyzing a dataset in which there are one or more independent variables which determine an outcome. The outcome is estimated with a dichotomous variable (in which there are only two possible outcomes). In logistic regression, the dependent variable is binary or dichotomous meaning that it only contains data coded as 1 (such as 'True', 'Yes') or 0 ('False', 'No'). If the dependent variable is continuous, it can be dichotomized at some logically meaningful cut point. The goal of logistic regression is to find the best fitting and logically reasonable model to describe the relationship between an outcome (dependent or response variable) and a set of independent (predictor or explanatory) variables [49]. Moreover, unlike the classic linear regression where the parameters are computed using the least squares method, logistic regression estimates the parameters using the likelihood ratio. In this way, the response variable (predicted) is a function of the likelihood that a particular observation (individual) will be in one of the two categories of the dichotomy.

In the present study, since the dependent variable was binary (dichotomous), the function of the binary logistic regression was:

$$f(Z) = \frac{e^z}{1+e^z} = \frac{1}{1+e^{-z}} \tag{1}$$

where Z is the input variable and $f(Z)$ its outcome. One of the advantages of this function is that the input variable takes positive and negative values, whereas the outcome $f(Z)$ ranges between 0 and 1. More analytically, the variable Z represents the combined influence of a set of variables, while $f(Z)$ defines the likelihood of a specific outcome resulting from this action.

In addition, variable Z expresses the measure of the overall contribution of all participating independent variables to the model and is defined as:

$$z = \beta_0 + \beta_1 X_1 + \beta_2 X_2 + \cdots + \beta_k X_k \tag{2}$$

where, $\beta_0$. is the intercept of the regression line and $\beta_i$ the coefficients of the independent variables, expressing the contribution of each variable.

When a coefficient takes a positive value, the explanatory variable increases the likelihood of a successful outcome (i.e., the realization of the event). Conversely, a negative coefficient value means that the variable decreases the likelihood of the outcome. In addition, a high value of the coefficient would signify that the independent variable significantly affects the likelihood of the realization of the event, whereas a low value would denote a small effect of the independent variable on the likelihood of having the relevant result.

## 4. Results

First, the results of descriptive statistics concerning the sociodemographic characteristics of the students participating in the survey are presented. The next part of the results reveals students' preferences for different energy types, their views on reasons for adopting renewables, and their daily environmental habits. Finally, the logistic model that emerged from the application of logistic regression is fully described and explained.

### 4.1. Sociodemographic Characteristics

According to Figure 1a, 53.7% of the participants were male and 46.3% female. As for their year of study, Figure 1b shows that the second- and third-year of study presented lower participation, while first, fourth, and fifth years had greater percentages.

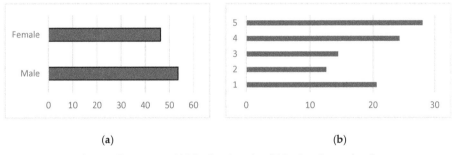

(a)  (b)

**Figure 1.** Percentages of (a) Students' gender; (b) Students' year of study.

The students' family background in terms of parental occupational and educational level were also examined. As given in Figure 2a, a significant share of the students' fathers was employed in the public (26.6%) and private (22%) sector. Likewise, as it appears in Figure 2b, a similar proportion of the participants' mothers were also employed in the public and private sectors (24.8% and 21.5%, respectively). Meanwhile, a substantial percentage of mothers (23.8%) were full-time housewives without being engaged in paid work. Finally, only 11.7% and 17.2% of parents were unemployed and pensioners, respectively.

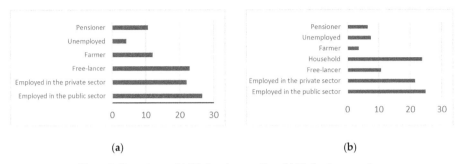

**Figure 2.** Percentages of (**a**) Fathers' occupation; (**b**) Mother's occupation.

With regard to parental education, overall students' parents exhibited a high educational level. A considerable proportion of fathers (36.9%) and mothers (45.3%) were higher education graduates, whereas the percentage of parents having attended merely compulsory education was notably lower (Figure 3a,b).

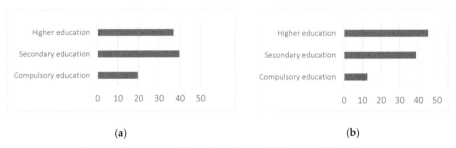

**Figure 3.** Percentages of Educational level of (**a**) father and (**b**) mother.

### 4.2. Descriptive Statistics

The surveyed students were asked whether they were willing to invest in renewable energy in the future. Results depicted in Figure 4 show that the clear majority of students, by 85%, was inclined to make a RE investment in their later life.

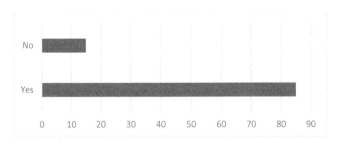

**Figure 4.** Percentages of students' willingness to invest in renewable energy sources (RES).

Afterwards, the students evaluated different energy production technologies based on the types they preferred to be further developed in Greece. For this evaluation, the students were given a 10-point scale (where 1 stands for "to be less developed" and 10 for "to be more developed") on which they marked their preference. As Figure 5 depicts, approximately seven out of ten students supported the development of solar energy and almost six out of ten supported wind energy. Conversely, only about one out of ten students favored nuclear fuels or coal combustion.

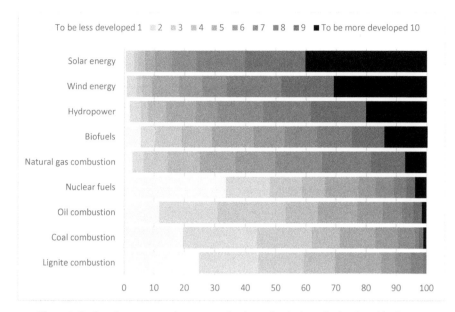

**Figure 5.** Students' assessment of energy production technologies to be developed in Greece.

Afterwards, students were asked to what degree they agreed or disagreed with various reasons for adopting renewable energy. Specifically, the students evaluated, using a five-point scale ranging from "strongly disagree" to "strongly agree", a variety of reasons for installing an RE system, such as wind energy or photovoltaics. According to our results (Figure 6), approximately eight out of ten students perceived that reduced pollution levels, country's increased energy independence, and improved air quality were the most important reasons for installing a renewable energy system.

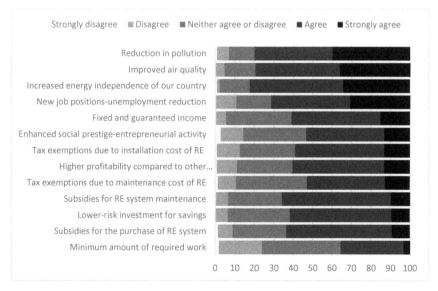

**Figure 6.** Students' degree of agreement with reasons for adopting RE systems.

To discover respondents' environmental attitudes, students evaluated their daily practices and habits on a five-point scale. Results given in Figure 7 indicated that relative to transportation, approximately eight out of ten students were willing to use the bicycle or cover short distances on foot to lower their individual environmental impact. As for energy saving, about eight out of ten students were willing to switch off the lights when leaving a room or opt for energy-efficient light bulbs and to turn off the tap while brushing teeth or shaving. Yet, the examination of their eating habits showed that only three out of ten participants were willing to reduce the consumption of meat and cured meat products for the sake of the environment.

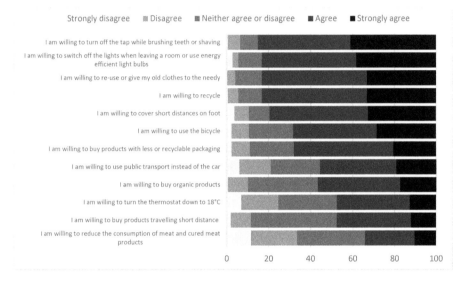

**Figure 7.** Students' daily environmental practices and habits.

*4.3. Logistic Regression Model*

Next, to predict "Students' willingness or unwillingness to invest in RE", Logistic Regression was conducted. The dependent variable was the "Students' willingness or unwillingness to invest in RE" (V1) and as independent variables the factor scores resulted from a factor analysis to the multivariates "Students' preference for energy sources to be developed" (V2), "Reasons for RE adoption" (V3), and "Participants' daily habits" (V4) were used. Factor analysis has been performed in previous work [26] and its outcomes are used here as input variables in the logistic regression model. The factors of the three multivariate variables are described below.

The multivariate "Students' preference for energy sources to be developed" (V2), gave three factors with significant loadings (V2_1, V2_2, V2_3). Factor V2_1 can be referred to as "Renewable sources and natural gas" since it includes the renewable energy types and natural gas ("Wind energy", "Hydropower", "Solar energy", and "Natural gas"). The second factor (V2_2) can be termed "Conventional energy types" because it involves explicitly fossil fuel-based energy technologies such as "Coal combustion", "Lignite combustion", and "Oil combustion". The third factor V2_3 can be referred to as "Nuclear fuels and biofuels" since it contains only these two energy sources.

The second multivariate "Reasons for RE adoption" (V3) examined participants' views on reasons for adopting renewables. The factor analysis resulted in five important factors (V3_1, V3_2, V3_3, V3_4, V3_5). Factor V3_1 can be termed "Environmentally- and energy independence-related reasons" because it contains the variables "Improved air quality", "Increased energy independence of our country" and "Reduction in pollution". The second factor (V3_2) can be referred to as "Reasons related to financial motives and minimum work" as it includes the variables "Subsidies for the purchase of the

RE system", "Subsidies for the maintenance of the RE system", "Fixed and guaranteed income", and "Minimum amount of work". The third factor (V3_3) contains the variables "Lower-risk investment for savings" and "Higher profitability compared to other investments" and thus can be called "Investment reasons". The fourth factor (V3_4) contains distinctly reasons related to tax exemptions that adopters are entitled to when they install an RES system ("Tax exemptions due to installation cost of RE" and "Tax exemptions due to maintenance cost of RE") and thereby V3_4 can be termed "Tax exemptions". The fifth and last factor (V3_5) can be termed "Socially- and employment-related reasons" because under this factor the variables "Enhanced social prestige-entrepreneurial activity", and "New job positions—unemployment reduction" were loaded.

The multivariate "Participants' daily habits" (V4) gave three factors. The first factor (V4_1) can be termed "Energy/water saving and recycling/reusing habits" since it includes the variables relating to energy/water saving and recycling habits. More specifically, these were "I am willing to switch off the lights when leaving a room or use energy-efficient light bulbs", "I am willing to recycle", "I am willing to re-use or give my old clothes to the needy" and "I am willing to turn off the tap while brushing teeth or shaving". The second factor (V4_2) can be referred to as "Transport and heating habits" since it contains variables concerning transport and heating choices which were "I am willing to use the bicycle", "I am willing to cover short distances on foot" and "I am willing to turn the thermostat down to 18 °C". Finally, since the third factor (V4_3) includes variables which relate to participants' behavior as consumers and their willingness to use public transport it can be termed "Consumer habits and willingness to use public transport". More specifically, the included variables are "I am willing to buy products travelling short distance", "I am willing to buy organic products", "I am willing to reduce the consumption of meat and cured meat products", "Instead of the car I am willing to use public transport", and "I am willing to buy products with less or recyclable packaging".

The scores of the above described factors were used to perform the logistic regression analysis and derive the model of Equation (2), with the students' willingness to invest in renewable energy as dependent variable. All factors (independent variables) were included in the analysis and the stepwise procedure with forward selection of variables was implemented to find out the significant ones. Table 1 shows the output of the logistic regression analysis related to the model fit tests and statistics.

Table 1. Logistic regression results—model fit indices.

| | | Omnibus Tests of Model Coefficients | | |
| --- | --- | --- | --- | --- |
| | | Chi-Square | Degrees of Freedom | Significance |
| Step 3 | Step | 4.850 | 1 | 0.028 |
| | Block | 29.925 | 3 | 0.000 |
| | Model | **29.925** | 3 | **0.000** |
| **Model Summary** | | | | |
| | | −2 Log likelihood | Cox & Snell R Square | Nagelkerke $R^2$ |
| | Step 3 | 150.647 [a] | 0.130 | 0.229 |
| | | Hosmer and Lemeshow Test | | |
| | | Chi-square | Degrees of Freedom | Significance |
| | Step 3 | 11.528 | 8 | 0.174 |

[a] Estimation terminated at iteration number five because parameter estimates changed by less than 0.05.

The Omnibus test of model coefficients gives a Chi-Square of 29.925 on 3 df, significant beyond 0.001, rejecting the null hypothesis that adding the variables to the model has not significantly increased our ability to predict the decisions made by the Students' (willingness or unwillingness) to invest in RE.

Under "Model Summary" (Table 1) we see that the −2 Log likelihood statistic is 150.647. This is the value that was compared to the −2 Log likelihood for the null model in the Omnibus test of model coefficients and resulted in significant predictability for the model. The $R^2$ values (Cox & Snell and

Nagelkerke's) are approximations of how much variation in the outcome is explained by the model. Nagelkerke's $R^2$ value indicates that the model explains roughly 23% of the variation in the outcome.

The Hosmer & Lemeshow test demonstrates the goodness-of-fit of the model, since the value of Chi-Square = 11.528 corresponds to a statistical significance greater than 0.05.

Next, the classification table (Table 2) compares the observed willingness of the students to invest in RE with that predicted by the model. The overall percentage of correct prediction is almost 90% (observations correctly classified), indicating a very good performance of the model.

Table 2. Classification table.

|        | Observed           |     | Predicted   |    | Percentage Correct |
|--------|--------------------|-----|-------------|----|--------------------|
|        |                    |     | Invest in RE|    |                    |
|        |                    |     | Yes         | No |                    |
| Step 3 | Invest in RE       | Yes | 181         | 1  | 99.5               |
|        |                    | No  | 26          | 6  | 18.8               |
|        | Overall Percentage |     |             |    | 87.4               |

The results showed that only three variables were significant, that is V2_3, V3_3, and V4_1. Accordingly, Equation (3) took the following form:

$$V1 = 2.097 + 0.500\ V2\_3* + 0.599\ V3\_3* + 0.840\ V4\_1* \quad (3)$$

where V1 is the dependent variable and V2_3 - V3_3 - V4_1 are the independent variables and (*) indicates the statistically significant beta coefficients at $p < 0.001$ level.

Table 3 presents the regression coefficients (B), the standard error (S.E.) of each coefficient, the Wald statistic (for the statistical significance testing), and the odds ratio (Exp (B)) for each variable in the model. All regression coefficients are statistically significant (at 0.01 level, except for V2_3 which is significant at 0.05 level) and positive, indicating that increasing influence of all variables is associated with increased odds of willingness to invest in RE. The 0.607 odds ratio for V2_3 indicates that the odds of investing are more than 60% for each one-point increase in V2_3 score. That is, for each one-point increase of V2_3, there is a 60.7% increase of the odds that the student will invest in RE. Similarly, for each one-point increase in the scores of V3_3 and V4_1, an increase of 55% and 43.2%, respectively, is expected in the odds that the student will invest in RE.

Table 3. Statistical significance of the variables in the model.

| Model Variables | B     | S.E.  | Wald   | Significance | Exp(B) |
|-----------------|-------|-------|--------|--------------|--------|
| V2_3            | 0.500 | 0.235 | 4.532  | 0.033        | 0.607  |
| V3_3            | 0.599 | 0.218 | 7.529  | 0.006        | 0.550  |
| V4_1            | 0.840 | 0.222 | 14.281 | 0.000        | 0.432  |
| Constant        | 2.097 | 0.246 | 72.849 | 0.000        | 0.123  |

V2_3: "Nuclear fuels and biofuels" to be developed
V3_3: "Investment reasons" (risk and profitability)
V4_1: "Energy/water saving and recycling/reusing habits"

## 5. Discussion

The purpose of this study was to analyze the factors that influence the willingness of environmental students to make investments in renewables. To that end, logistic regression analysis computed various factors to discover which of them affected participants' willingness to invest the most, and "Energy/water saving and recycling/reusing habits" was identified as a factor with great influence. In other words, the likelihood to invest in renewable energy increases when respondents adopt pro-environmental behaviors, particularly in terms of energy/water saving and recycling. This finding resonates with the observations of theoretical social psychology which indicated that individuals with positive environmental attitudes are likely to adopt a responsible behavior that involves a lower environmental impact [50], which could possibly involve investments in technologies that produce clean energy. Hence, our study confirms the findings of previous studies showing that citizens who install renewables were mainly motivated by their environmental awareness and positive environmental attitudes [23,27–31].

Additionally, the analysis revealed that "Lower-risk investment for savings" and "Higher profitability compared to other investments" was another investment affecting factor. Therefore, the likelihood to invest in renewables will increase when the involved investment risk is low and when the expected profitability is higher than that of other types of investments. Moreover, in view of this finding, it is possible to assume that students may have acknowledged that renewable investments can be long-term and perhaps more secure than other investments activities such as buying bonds and shares.

The third factor, which according to our analysis was important to investments, was "Nuclear fuels and biofuels", showing that investments will increase if these energy types are further developed in Greece. Here, the term 'biofuels' refers to fuels which are produced from biomass; however, as with other agricultural procedures, biofuel feedstock production can have impacts on sustainability which are context-specific [51], with the most severe impacts being greenhouse gas emissions, changes in land-use and water-use, as well as water and air pollution [52,53]. These impacts can result from the cultivation of plants and crops which are necessary for the production of biofuels which is often criticized as unsustainable [54]. On the other hand, nuclear power is commonly regarded as exceedingly hazardous, thereby causing negative public reactions [55,56]. Despite the drawbacks of these two energy types, both nuclear power and biofuels comprise technologies with low carbon emissions which are less implemented in Greece and this can perhaps explain why 'Nuclear fuels and biofuels' emerged from the analysis as an important investment factor. To understand students' mindset in relation to these contentious energy production technologies, qualitative-oriented studies would be effective as they could reveal how environmental students perceive energy technologies which often receive skepticism.

According to our analysis, "Tax exemptions" and "Reasons related to financial motives and minimum work" were found to be unimportant to investments and thus are of the greatest interest to discuss. The indicated unimportant role of tax incentives is in sharp contrast with other studies which have shown that tax incentives are able to attract investments [57,58]. This implies that the current taxation for investments in renewable energies in Greece is unfavorable and as such it does not comprise a factor that will increase small-scale investments. Likewise, it is interesting that the variables "Subsidies for the purchase of the RES system", "Subsidies for the maintenance of the RES system", and "Fixed and guaranteed income" were not indicated as significant investment factors. Again, this may suggest that the respondents were not satisfied with the provided subsidies or with the income that microgenerators receive for the produced electricity (feed-in-tariffs). However, this is inconsistent with the findings of studies conducted in other countries showing that subsidies and feed-in-tariffs played a significant role in attracting small-scale investments [23,32,59].

In the present study, students in their overwhelming majority reported their intention to invest in renewables in the future, thereby resonating with previous findings which have also revealed willingness among university students to make renewable investments in the future [43,47,48]. Furthermore, the willingness that was recorded in our survey could have been related to the participants' sociodemographic characteristics, such as their education level. In specific, although the surveyed students have not graduated yet, they could be regarded as having a high level of education since they have been studying at the university for at least one year. In this light, their high educational level could have positively influenced their willingness to invest and this interpretation is validated by other studies showing that highly educated individuals are more positively inclined towards renewable energy investments [29,34].

As mentioned earlier, the age of potential investors appears to be influential to individuals' intention to invest without, however, knowing whether this is a positive or negative influence since the findings of the relevant literature are conflicting. Our sample explicitly consisted of undergraduate students and thus most respondents were aged between 19 and 23, while they clearly expressed their willingness to invest. Hence, our results suggest that young age is positively related to investment willingness, thereby confirming similar former findings [24].

Moreover, students' stated preference for solar and wind power in specific resonates with other studies conducted in Greece [34,60–62] as well as studies performed in other countries [63,64]. With regard to Greece, solar and wind are not only implemented to a higher degree in comparison to other renewables, but also their installations are located in different areas and in more apparent places than the plants of other RES such as biomass and hydropower [65]. From this perspective, it is reasonable to assume that respondents favored mostly solar and wind power because they were more familiar with these two renewable types.

In contrast to their support for renewable types, participants expressed a remarkably limited preference for fossil fuel-based energy production technologies. Again, their background of environmental studies could explain their little support for conventional fuels. That is, students' knowledge of the detrimental impacts of this technology on the environment is likely to have played a role in forming negative attitudes to fossils and raising their awareness about renewable energies.

Another interesting finding was that respondents rated the energy independence of the country as the most important reason for installing renewable energy. To contextualize this finding, it is important to note that, currently, Greece imports fossil fuels to meet the greatest part of its energy needs; however, fuel imports can have negative effects on the economy of a country because fuel prices are subjected to abrupt increases due to geopolitical crises. In this context, students must have been conscious about the economic impacts of fuel imports and must have acknowledged that higher installed capacities of renewable energies can help the country to become energy independent and secure its economy from fuel price fluctuations. Respondents may also have known that Greece could become energy independent because it presents an impressive renewable energy potential and is greatly advantageous regarding its wind and solar energy potential [66].

In terms of their environmental behavior, respondents have mostly adopted energy and water saving habits probably because they know through their studies that all daily habits involve a certain environmental impact regardless of how simple they may seem. Nevertheless, the participants did not express the same awareness about the consumption of meat and seemed to be unwilling to reduce its consumption for the sake of the environment. Interestingly, this finding is consistent with that of a previous study [62] which also detected reluctance to decrease meat intake among secondary school students in Northern Greece. The reported unwillingness in both cases could be related to the absence of large meat industries in Greece and to the fact that the country imports the biggest part of beef, while only a small amount is produced locally. This means that the contribution of the meat industry's methane emissions to environmental issues has not been discussed sufficiently, suggesting that the respondents were probably ignoring that the frequent consumption of meat involves a negative impact on the environment. However, since they have clearly reported their intention to be careful about their environmental behavior in terms of energy and water consumption, it can be stated that the students would possibly also become mindful of their diet habits if they knew more about the impact of the meat industry on the environment.

## 6. Conclusions

Knowing the factors which affect investment decisions is a crucial step in order to increase the amount of investments made in renewables. With regard to our findings, it can be inferred that investments increase when individuals adopt pro-environmental behaviors that focus on energy and water saving and when investments in renewables are safer and more profitable compared to other types of investments. In addition, the development of biofuels and nuclear fuels could also increase investments as their development was also indicated as a significant factor in the final logistic regression model.

Since willingness to invest was irrespective of certain variables, it would be highly relevant to improve these areas in order to trigger investments in renewables. In particular, since tax exemptions were not identified as investment factors, further tax exemptions or tax reductions should be introduced for individuals who purchase renewable energy systems or perform maintenance on their installations. In addition, our analysis has shown that the provided financial incentives and the income from renewable investments were not positively related to respondents' willingness to invest, suggesting that the current subsidy system and the provided income from investments are not effective in attracting investments. To reverse this negative trend, a more generous subsidy system should be applied which would specifically finance the purchase and maintenance of small-scale installations, while the quota of the income from investments in renewables should be increased.

Moreover, our study has highlighted certain areas for future research. Since participants expressed a greater preference for wind and solar energy than other renewable types, the reasons behind the limited support for other renewables ought to be studied. In our discussion, we have attributed it to their higher familiarity with solar and wind as these consist the most implemented renewables in Greece. However, future research work should investigate students' knowledge of and attitudes towards hydropower, biomass, and geothermal power.

Despite the fact that the present study was conducted during a time of economic recession, the overwhelming majority of students was found willing to invest in environmentally friendly technologies. It would be of great interest to perform a similar study in times of more favorable economic conditions and compare the results of both studies to indicate whether financial difficulties were positively related to participants' increased investment willingness.

As previously discussed, respondents' reluctance to reduce meat for the protection of the environment could be a result of the lack of large cattle farms in Greece and the subsequent lack of debates on methane emissions' contribution to environmental issues. Although this appears to be a reasonable explanation, its validity should be confirmed by further studies.

Moreover, to ensure that the students' investment willingness will be turned into actual investment in the years to come, policymakers and developers should create a favorable investment environment that will meet the expectations and needs of young individuals who wish to invest in renewable energy. For this special category of potential investors, loans at low interest rates could be granted, thereby enabling them to invest in their own plant or in an energy project.

Finally, a limitation of our study is that it has examined only the investment willingness of environmental students as these were considered the most suitable respondents given the purposes of our study. However, it is equally important to examine the willingness of students majoring in other disciplines in Greece and elsewhere, because they are also potential investors and their intention to invest in renewable energy ought to be examined.

**Author Contributions:** E.K. collected the data and reviewed the relevant literature. S.G. performed statistical analysis. G.T. and S.G. prepared the methodology. E.K. and G.T. wrote the original draft. S.G. reviewed and edited the manuscript. G.T. supervised and coordinated the work.

**Funding:** This research received no external funding.

**Conflicts of Interest:** The authors declare no conflict of interest.

## References

1. REN21. *Renewables 2019 Global Status Report*; REN21: Paris, France, 2019; ISBN 978-3-9818911-7-1.
2. Ioannou, K.; Tsantopoulos, G.; Arabatzis, G.; Andreopoulou, Z.; Zafeiriou, E. A Spatial Decision Support System Framework for the Evaluation of Biomass Energy Production Locations: Case Study in the Regional Unit of Drama, Greece. *Sustainability* **2018**, *10*, 531. [CrossRef]
3. International Energy Agency. *Energy Policies of IEA Countries-Greece 2017 Review*; International Energy Agency: Paris, France, 2017.

4. Papadopoulos, A.M.; Karteris, M.M. An assessment of the Greek incentives scheme for photovoltaics. *Energy Policy* **2009**, *37*, 1945–1952. [CrossRef]
5. Lazarou, S.; Pyrgioti, E.; Agoris, D. The latest Greek statute laws and its consequences to the Greek renewable energy source market. *Energy Policy* **2007**, *35*, 4009–4017. [CrossRef]
6. Karteris, M.; Papadopoulos, A.M. Legislative framework for photovoltaics in Greece: A review of the sector's development. *Energy Policy* **2013**, *55*, 296–304. [CrossRef]
7. Forouli, A.; Gkonis, N.; Nikas, A.; Siskos, E.; Doukas, H.; Tourkolias, C. Energy efficiency promotion in Greece in light of risk: Evaluating policies as portfolio assets. *Energy* **2019**, *170*, 818–831. [CrossRef]
8. Bergek, A.; Mignon, I.; Sundberg, G. Who invests in renewable electricity production? Empirical evidence and suggestions for further research. *Energy Policy* **2013**, *56*, 568–581. [CrossRef]
9. Gamel, J.; Menrad, K.; Decker, T. Which factors influence retail investors' attitudes towards investments in renewable energies? *Sustain. Prod. Consum.* **2017**, *12*, 90–103. [CrossRef]
10. Claudy, M.C.; Michelsen, C.; O'Driscoll, A.; Mullen, M.R. Consumer awareness in the adoption of microgeneration technologies. *Renew. Sustain. Energy Rev.* **2010**, *14*, 2154–2160. [CrossRef]
11. Lee, C.-Y.; Heo, H. Estimating willingness to pay for renewable energy in South Korea using the contingent valuation method. *Energy Policy* **2016**, *94*, 150–156. [CrossRef]
12. Ma, C.; Rogers, A.A.; Kragt, M.E.; Zhang, F.; Polyakov, M.; Gibson, F.; Chalak, M.; Pandit, R.; Tapsuwan, S. Consumers' willingness to pay for renewable energy: A meta-regression analysis. *Resour. Energy Econ.* **2015**, *42*, 93–109. [CrossRef]
13. Soon, J.-J.; Ahmad, S.-A. Willingly or grudgingly? A meta-analysis on the willingness-to-pay for renewable energy use. *Renew. Sustain. Energy Rev.* **2015**, *44*, 877–887. [CrossRef]
14. Koto, P.S.; Yiridoe, E.K. Expected willingness to pay for wind energy in Atlantic Canada. *Energy Policy* **2019**, *129*, 80–88. [CrossRef]
15. Sun, C.; Yuan, X.; Xu, M. The public perceptions and willingness to pay: From the perspective of the smog crisis in China. *J. Clean. Prod.* **2016**, *112*, 1635–1644. [CrossRef]
16. Tan, R.; Lin, B. Public perception of new energy vehicles: Evidence from willingness to pay for new energy bus fares in China. *Energy Policy* **2019**, *130*, 347–354. [CrossRef]
17. Su, W.; Liu, M.; Zeng, S.; Štreimikienė, D.; Baležentis, T.; Ališauskaitė-Šeškienė, I. Valuating renewable microgeneration technologies in Lithuanian households: A study on willingness to pay. *J. Clean. Prod.* **2018**, *191*, 318–329. [CrossRef]
18. Kostakis, I.; Sardianou, E. Which factors affect the willingness of tourists to pay for renewable energy? *Renew. Energy* **2012**, *38*, 169–172. [CrossRef]
19. Zografakis, N.; Gillas, K.; Pollaki, A.; Profylienou, M.; Bounialetou, F.; Tsagarakis, K.P. Assessment of practices and technologies of energy saving and renewable energy sources in hotels in Crete. *Renew. Energy* **2011**, *36*, 1323–1328. [CrossRef]
20. Curtin, J.; McInerney, C.; Gallachóir, B.Ó.; Salm, S. Energizing local communities—What motivates Irish citizens to invest in distributed renewables? *Energy Res. Soc. Sci.* **2019**, *48*, 177–188. [CrossRef]
21. Hai, M.A. Rethinking the social acceptance of solar energy: Exploring "states of willingness" in Finland. *Energy Res. Soc. Sci.* **2019**, *51*, 96–106. [CrossRef]
22. Ebers Broughel, A.; Hampl, N. Community financing of renewable energy projects in Austria and Switzerland: Profiles of potential investors. *Energy Policy* **2018**, *123*, 722–736. [CrossRef]
23. Braito, M.; Flint, C.; Muhar, A.; Penker, M.; Vogel, S. Individual and collective socio-psychological patterns of photovoltaic investment under diverging policy regimes of Austria and Italy. *Energy Policy* **2017**, *109*, 141–153. [CrossRef]
24. Masini, A.; Menichetti, E. Investment decisions in the renewable energy sector: An analysis of non-financial drivers. *Technol. Forecast. Soc. Chang.* **2013**, *80*, 510–524. [CrossRef]
25. Papadopoulou, S.-D.; Kalaitzoglou, N.; Psarra, M.; Lefkeli, S.; Karasmanaki, E.; Tsantopoulos, G. Addressing Energy Poverty through Transitioning to a Carbon-Free Environment. *Sustainability* **2019**, *11*, 2634. [CrossRef]
26. Karasmanaki, E.; Tsantopoulos, G. Exploring future scientists' awareness about and attitudes towards renewable energy sources. *Energy Policy* **2019**, *131*, 111–119. [CrossRef]
27. Parkins, J.R.; Rollins, C.; Anders, S.; Comeau, L. Predicting intention to adopt solar technology in Canada: The role of knowledge, public engagement, and visibility. *Energy Policy* **2018**, *114*, 114–122. [CrossRef]

28. Strazzera, E.; Statzu, V. Fostering photovoltaic technologies in Mediterranean cities: Consumers' demand and social acceptance. *Renew. Energy* **2017**, *102*, 361–371. [CrossRef]
29. Michelsen, C.C.; Madlener, R. Switching from fossil fuel to renewables in residential heating systems: An empirical study of homeowners' decisions in Germany. *Energy Policy* **2016**, *89*, 95–105. [CrossRef]
30. Palm, J.; Tengvard, M. Motives for and barriers to household adoption of small-scale production of electricity: Examples from Sweden. *Sustain. Sci. Pract. Policy* **2011**, *7*, 6–15. [CrossRef]
31. Zhai, P.; Williams, E.D. Analyzing consumer acceptance of photovoltaics (PV) using fuzzy logic model. *Renew. Energy* **2012**, *41*, 350–357. [CrossRef]
32. Ozcan, M. Assessment of renewable energy incentive system from investors' perspective. *Renew. Energy* **2014**, *71*, 425–432. [CrossRef]
33. Ek, K.; Persson, L.; Johansson, M.; Waldo, Å. Location of Swedish wind power—Random or not? A quantitative analysis of differences in installed wind power capacity across Swedish municipalities. *Energy Policy* **2013**, *58*, 135–141. [CrossRef]
34. Tsantopoulos, G.; Arabatzis, G.; Tampakis, S. Public attitudes towards photovoltaic developments: Case study from Greece. *Energy Policy* **2014**, *71*, 94–106. [CrossRef]
35. Frederiks, E.R.; Stenner, K.; Hobman, E.V. Household energy use: Applying behavioural economics to understand consumer decision-making and behaviour. *Renew. Sustain. Energy Rev.* **2015**, *41*, 1385–1394. [CrossRef]
36. Leete, S.; Xu, J.; Wheeler, D. Investment barriers and incentives for marine renewable energy in the UK: An analysis of investor preferences. *Energy Policy* **2013**, *60*, 866–875. [CrossRef]
37. Jung, N.; Moula, M.E.; Fang, T.; Hamdy, M.; Lahdelma, R. Social acceptance of renewable energy technologies for buildings in the Helsinki Metropolitan Area of Finland. *Renew. Energy* **2016**, *99*, 813–824. [CrossRef]
38. Bollinger, B.; Gillingham, K. Peer Effects in the Diffusion of Solar Photovoltaic Panels. *Mark. Sci.* **2012**, *31*, 900–912. [CrossRef]
39. Carlisle, J.E.; Solan, D.; Kane, S.L.; Joe, J. Utility-scale solar and public attitudes toward siting: A critical examination of proximity. *Land Use Policy* **2016**, *58*, 491–501. [CrossRef]
40. Tabi, A.; Wüstenhagen, R. Keep it local and fish-friendly: Social acceptance of hydropower projects in Switzerland. *Renew. Sustain. Energy Rev.* **2017**, *68*, 763–773. [CrossRef]
41. Karatepe, Y.; Neşe, S.V.; Keçebaş, A.; Yumurtacı, M. The levels of awareness about the renewable energy sources of university students in Turkey. *Renew. Energy* **2012**, *44*, 174–179. [CrossRef]
42. Yazdanpanah, M.; Komendantova, N.; Shirazi, Z.N.; Linnerooth-Bayer, J. Green or in between? Examining youth perceptions of renewable energy in Iran. *Energy Res. Soc. Sci.* **2015**, *8*, 78–85. [CrossRef]
43. Qu, M.; Ahponen, P.; Tahvanainen, L.; Gritten, D.; Mola-Yudego, B.; Pelkonen, P. Chinese university students' knowledge and attitudes regarding forest bio-energy. *Renew. Sustain. Energy Rev.* **2011**, *15*, 3649–3657. [CrossRef]
44. Cotton, D.; Shiel, C.; Paço, A. Energy saving on campus: A comparison of students' attitudes and reported behaviours in the UK and Portugal. *J. Clean. Prod.* **2016**, *129*, 586–595. [CrossRef]
45. Ahamad, N.R.; Ariffin, M. Assessment of knowledge, attitude and practice towards sustainable consumption among university students in Selangor, Malaysia. *Sustain. Prod. Consum.* **2018**, *16*, 88–98. [CrossRef]
46. Alawin, A.A.; Rahmeh, T.A.; Jaber, J.O.; Loubani, S.; Dalu, S.A.; Awad, W.; Dalabih, A. Renewable energy education in engineering schools in Jordan: Existing courses and level of awareness of senior students. *Renew. Sustain. Energy Rev.* **2016**, *65*, 308–318. [CrossRef]
47. Ozil, E.; Ugursal, V.I.; Akbulut, U.; Ozpinar, A. Renewable Energy and Environmental Awareness and Opinions: A Survey of University Students in Canada, Romania, and Turkey. *Int. J. Green Energy* **2008**, *5*, 174–188. [CrossRef]
48. Assali, A.; Khatib, T.; Najjar, A. Renewable energy awareness among future generation of Palestine. *Renew. Energy* **2019**, *136*, 254–263. [CrossRef]
49. Hosmer, D.W.; Hosmer, T.; le Cessie, S.; Lemeshow, S. A comparison of goodness-of-fit tests for the logistic regression model. *Stat. Med.* **1997**, *16*, 965–980. [CrossRef]
50. Sapci, O.; Considine, T. The link between environmental attitudes and energy consumption behavior. *J. Behav. Exp. Econ.* **2014**, *52*, 29–34. [CrossRef]
51. Graham, J.B.; Stephenson, J.R.; Smith, I.J. Public perceptions of wind energy developments: Case studies from New Zealand. *Energy Policy* **2009**, *37*, 3348–3357. [CrossRef]

52. Shonnard, D.R.; Klemetsrud, B.; Sacramento-Rivero, J.; Navarro-Pineda, F.; Hilbert, J.; Handler, R.; Suppen, N.; Donovan, R.P. A Review of Environmental Life Cycle Assessments of Liquid Transportation Biofuels in the Pan American Region. *Environ. Manag.* **2015**, *56*, 1356–1376. [CrossRef]
53. Filoso, S.; do Carmo, J.B.; Mardegan, S.F.; Lins, S.R.M.; Gomes, T.F.; Martinelli, L.A. Reassessing the environmental impacts of sugarcane ethanol production in Brazil to help meet sustainability goals. *Renew. Sustain. Energy Rev.* **2015**, *52*, 1847–1856. [CrossRef]
54. Correa, D.F.; Beyer, H.L.; Fargione, J.E.; Hill, J.D.; Possingham, H.P.; Thomas-Hall, S.R.; Schenk, P.M. Towards the implementation of sustainable biofuel production systems. *Renew. Sustain. Energy Rev.* **2019**, *107*, 250–263. [CrossRef]
55. Kim, Y.; Kim, M.; Kim, W. Effect of the Fukushima nuclear disaster on global public acceptance of nuclear energy. *Energy Policy* **2013**, *61*, 822–828. [CrossRef]
56. Thomas, S. What will the Fukushima disaster change? *Energy Policy* **2012**, *45*, 12–17. [CrossRef]
57. Onuoha, I.J.; Aliagha, G.U.; Rahman, M.S.A. Modelling the effects of green building incentives and green building skills on supply factors affecting green commercial property investment. *Renew. Sustain. Energy Rev.* **2018**, *90*, 814–823. [CrossRef]
58. Rabab Mudakkar, S.; Zaman, K.; Shakir, H.; Arif, M.; Naseem, I.; Naz, L. Determinants of energy consumption function in SAARC countries: Balancing the odds. *Renew. Sustain. Energy Rev.* **2013**, *28*, 566–574. [CrossRef]
59. Baharoon, D.A.; Rahman, H.A.; Fadhl, S.O. Publics' knowledge, attitudes and behavioral toward the use of solar energy in Yemen power sector. *Renew. Sustain. Energy Rev.* **2016**, *60*, 498–515. [CrossRef]
60. Zografakis, N.; Sifaki, E.; Pagalou, M.; Nikitaki, G.; Psarakis, V.; Tsagarakis, K.P. Assessment of public acceptance and willingness to pay for renewable energy sources in Crete. *Renew. Sustain. Energy Rev.* **2010**, *14*, 1088–1095. [CrossRef]
61. Liarakou, G.; Gavrilakis, C.; Flouri, E. Secondary School Teachers' Knowledge and Attitudes towards Renewable Energy Sources. *J. Sci. Educ. Technol.* **2009**, *18*, 120–129. [CrossRef]
62. Keramitsoglou, K.M. Exploring adolescents' knowledge, perceptions and attitudes towards Renewable Energy Sources: A colour choice approach. *Renew. Sustain. Energy Rev.* **2016**, *59*, 1159–1169. [CrossRef]
63. Çakirlar Altuntaş, E.; Turan, S.L. Awareness of secondary school students about renewable energy sources. *Renew. Energy* **2018**, *116*, 741–748. [CrossRef]
64. Zyadin, A.; Puhakka, A.; Ahponen, P.; Cronberg, T.; Pelkonen, P. School students' knowledge, perceptions, and attitudes toward renewable energy in Jordan. *Renew. Energy* **2012**, *45*, 78–85. [CrossRef]
65. Karytsas, S.; Theodoropoulou, H. Socioeconomic and demographic factors that influence publics' awareness on the different forms of renewable energy sources. *Renew. Energy* **2014**, *71*, 480–485. [CrossRef]
66. Kyriakopoulos, G.L.; Arabatzis, G.; Tsialis, P.; Ioannou, K. Electricity consumption and RES plants in Greece: Typologies of regional units. *Renew. Energy* **2018**, *127*, 134–144. [CrossRef]

© 2019 by the authors. Licensee MDPI, Basel, Switzerland. This article is an open access article distributed under the terms and conditions of the Creative Commons Attribution (CC BY) license (http://creativecommons.org/licenses/by/4.0/).

MDPI  
St. Alban-Anlage 66  
4052 Basel  
Switzerland  
Tel. +41 61 683 77 34  
Fax +41 61 302 89 18  
www.mdpi.com

*Sustainability* Editorial Office  
E-mail: sustainability@mdpi.com  
www.mdpi.com/journal/sustainability

Lightning Source UK Ltd.
Milton Keynes UK
UKHW052209190922
409095UK00002B/45